CYCLODEXTRIN CHEMISTRY

Preparation and Application

CYCLODEXTRIN CHEMISTRY

Preparation and Application

Zheng-Yu Jin

Jiangnan University, China

Chemical Industry Press

World Scientific

NEW JERSEY · LONDON · SINGAPORE · BEIJING · SHANGHAI · HONG KONG · TAIPEI · CHENNAI

Published by

World Scientific Publishing Co. Pte. Ltd.

5 Toh Tuck Link, Singapore 596224

USA office: 27 Warren Street, Suite 401-402, Hackensack, NJ 07601

UK office: 57 Shelton Street, Covent Garden, London WC2H 9HE

and

Chemical Industry Press
No. 13 Qingnianhu South Street
Dongcheng District
Beijing 100011
P.R. China

British Library Cataloguing-in-Publication Data
A catalogue record for this book is available from the British Library.

This edition is jointly published by World Scientific Publishing Co. Pte. Ltd. and Chemical Industry Press, P. R. China.

CYCLODEXTRIN CHEMISTRY
Preparation and Application

ISBN 978-981-4436-79-3

Typeset by Stallion Press
Email: enquiries@stallionpress.com

PREFACE

Cyclodextrins (sometimes called cycloamyloses) are a family of compounds made up of sugar molecules bound together in a ring (cyclic oligosaccharides). Cyclodextrins, produced from starch by means of enzymatic conversion, can form inclusion complexes with a considerable number of organic and inorganic compounds. The formation of inclusion complexes can markedly modify the physicochemical parameters of the guest molecule, adsorption capacity, polarity, hydrophobicity, etc. They are used in food, pharmaceutical, cosmetic and chemical industries, as well as in agriculture and environmental engineering, supermolecule and analytical chemistry.

Many developments have occurred in the world of cyclodextrins preparations, analyses, derivatives and applications since "cellulosine" was first described by Villier in 1891. Research in my group centers around the synthesis, analysis and application of branched and large cyclodextrins. The idea towards production of *Cyclodextrin Chemistry* germinated in 2007, and the actual writing started only when *Cyclodextrin Chemistry: Preparation and application (Chinese)* was published in 2009. I was pleased to see this book through to completion. *Cyclodextrin Chemistry* reviews processes, enzymes and analyses for preparing cyclodextrins and cyclodextrin derivatives, and their application in industrial and non-industrial areas. The big focus of this English edition was to do a general update, adding many new methods and topics. *Cyclodextrin Chemistry* will be a standard reference book for those working in the cyclodextrin area.

I am grateful to all the chapter authors for agreeing to be a part of this project. Many authors have drawn on their experiences with cyclodextrins. I wish to thank the authors of articles and books and their publishers for their permission to reproduce materials used here. Special thanks are extended to the following persons: Yao-Qi Tian and Xing Zhou for valuable discussions about the content of the book and assistance with editing; Xue-Ming Xu for inputs that helped determine content; Junrong Huang for word processing assistance.

Zheng-Yu Jin
Wuxi, Jiangsu, China
March 2012

ACKNOWLEDGMENT

This book skillfully brings together the work of many contributors who are experts in their respective cyclodextrin areas. It would not have been possible without their assistance and contributions. We are grateful to them for their participation and patience during the preparation of each chapter. We hope that readers will find this book useful. Our special thanks are also to our academic institutions, Jiangnan University and Shanxi University of Science and Technology for allowing us to dedicate our effort and time to the completion of this book and to our editors at Chemical Industry Press, Gang Wu and Yu-Qing Zhao, for supporting and coordinating the production of the book and their invaluable patience to answer our endless questions about the final proofing of the book. Overall, we could not finish such a task without the unconditional support of our families, so our most grateful thanks go to all of them.

Zheng-Yu Jin
Wuxi, Jiangsu, China
March 2012

CONTENTS

1

INTRODUCTION

Jun-Rong Huang*, Hai-Ning Zhuang† and Zheng-Yu Jin‡

*College of Life Science and Engineering,
Shaanxi University of Science and Technology,
Xi'an 710021, China
†School of Food Science and Technology,
Jiangnan University, Wuxi 214122, China
‡The State Key Laboratory of Food Science and Technology,
School of Food Science and Technology,
Jiangnan University, Wuxi 214122, China

Cyclodextrins (CDs) are cyclic oligosaccharides produced from the degradation of starch by cyclodextrin glucanotransferase (CGTase). The most common and commercially successful CDs are the α-, β- and γ-CDs which consist of 6, 7 and 8 glucose molecules, respectively. They were first discovered by Villier from the *Bacillus amylobacter* digest of potato starch [1]. The main interest in CDs lies in their ability to form inclusion complexes with several compounds [2]. They are ring- or torus-shaped molecules and posses a hydrophobic cavity and a hydrophilic exterior. The hydrophobicity of the cavity allows the CDs to associate with nonpolar organic molecules or portion of organic molecules to form inclusion complexes. One or two *guest* molecules can be entrapped by 1, 2 or 3 CDs [3]. CDs have been applied in the pharmaceutical, food, petroleum, paper, cosmetic, textile, etc. industries, and in analytical chemistry, agriculture and toilet articles [2, 4, 5].

1.1. History

In 1891, Villier isolated about 3 g of a material, which forms beautiful radiate crystals, from 1,000 g starch after digestion with *B. amylobacter*, which he named "cellulosine", later to be known as CD. According to other authors, Villiers probably used impure cultures and the CDs were produced by a *Bacillus macerans* contamination [2]. The composition of "cellulosine" was determined to be $(C_6H_{10}O_5)_2 \cdot 3H_2O$. It did not show reducing properties and was resistant against acidic hydrolysis [4].

The fundamentals of CD chemistry were laid down by Schardinger. In 1903, Schardinger published a report that digesting starch with a microorganism that survived the cooking process resulted in the formation of small amounts of two different crystalline products, dextrins A and B, which were described with regard to their lack of reducing power and seemed to be identical with the "cellulosines" of Villiers.

In 1904, Schardinger isolated a new organism capable of producing acetone and ethyl alcohol from sugar and starch-containing plant material. In 1911, he named the isolated microbe *B. macerans*, and reported that about 25%–30% of the starch could be converted to crystalline dextrins (with an additional larger amount of amorphous dextrins). Schardinger observed that the crystalline dextrins formed characteristic iodine adducts upon the addition of iodine–iodide solution. He named his crystalline products "crystallized dextrin α" and "crystallized dextrin β". The crystalline α-dextrin/iodine complex in thin layers is blue when damp and gray-green when dry, while the crystalline β-dextrins/iodine complex is brownish (red-brown) damp or dry [2, 4].

In the 1930s, Freudenberg and co-workers found that the crystalline Schardinger dextrins contain only α-1, 4-glycosidic linkages and are built from maltose units. In 1936, they postulated the cyclic structure of these crystalline dextrins [4, 6, 7]. In 1942, the structures of α- and β-CD were determined by X-ray crystallography [2]. In 1948, it was recognized that CDs can form inclusion complexes [2]. In 1953, Freudenberg, Camer and Plieinger obtained a patent using CD complexation which represented the application of CDs in drug formulations for protection from oxidation, enhancement of solubility, stabilizing volatile substance, etc. [4, 6].

γ-dextrin was not isolated until 1935 [2] and its X-ray structure was elucidated in 1948 [1]. The first indication of the existence of CDs comprising more than eight glycosyl units was published in 1948 by Freudenberg and Cramer [1]. In 1961, evidence for the natural existence of δ-, ζ-, ξ- and even η-CD (9–12 residues) was provided [2]. In 1965, French *et al.*, reported the first definitive evidence for the existence of large-ring cyclodextrins (LR-CDs) with a degree of polymerization (DP) from 9 to 13 [7]. The lab-scale preparation methods of CDs, their structure, physical and chemical properties, as well as their inclusion complex forming properties had been discovered by the end of 1960s. However, the CDs were very expensive at that time and were available only as fine chemicals [4, 6].

French published the first fundamental review in 1957 with the first misinformation on the toxicity of CDs, which deterred many scientists from developing CD-containing products for human use during the following 25 years [4, 6]. There was an explosion like increase in the number of CD-related publications after adequate toxicological studies proved that CDs were non-toxic from the 1970s onward. The first International Symposium on Cyclodextrins was held in 1981 and an International CD symposium has been organized every alternate year from 1984. The price of β-CD decreased from around $2,000 to several dollars per kilogram. The α- and γ-CDs, as well as several derivatives (hydroxypropyl-β-CD and γ-CD, randomly methylated α- and β-CDs, maltosyl-β-CD, acetylated CDs, etc.) are produced industrially [4, 6].

In 1986, Kobayashi *et al.* developed a preparation method for LR-CD mixtures and succeeded in isolating δ-CD (DP9) [7]. The crystal structure of δ-CD was characterized by Fujiwara *et al.* in 1990 [7]. Since it was difficult to experimentally distinguish the large CDs from branched CDs, the existence of the large CDs was confirmed by Ueda's group only during 1990s [1, 7]. They isolated and characterized LR-CDs with a DP from 9 to 21 [7]. Takaha *et al.* reported LR-CD with a DP up to 31 and their new synthesis in significant amounts using various glucanotranferase enzymes [7]. In addition, Machida *et al.* reported LR-CD mixtures with a DP from 22 to 45 [7], and the existence of even larger CDs with DP up to several hundreds of glycosyl units have been reported [1].

1.2. Nomenclature, Classification, Structure and Property

1.2.1. *Nomenclature*

"Cellulosine" was the first name used for CDs introduced by Villier. Later, Schardinger renamed the two isolated non-reducing crystalline compounds, dextrins A and B as "crystallized dextrin α" and "crystallized dextrin β". γ-Dextrin was introduced by Freudenberg and Jacobi. Because of the pioneering work of Schardinger, the CDs have often been denoted as "Schardinger dextrins" by the earlier researchers [1].

The term "cyclodextrin" has served as a general name for the α-, β- and γ-CDs for many years. However, the term cyclodextrin does not contain information on the nature of the intersaccharidic linkages. The use of semisystematic names for CDs was recommended by the Joint Commision on Biochemical Nomenclature in 1996 *"by citing the prefix cyclo, followed by the terms indicating the type of intersaccharidic linkages, the number of units and the termination '-ose'" (e.g., "malto" for α-1,4 linked glucose units; "hexa" for six).* The semisystematic names, such as cyclomaltohexaose for the CD consisting of six α-1,4-linked glycosyl units, have been used almost consistently as descriptors for the small CDs along with the Greek letter prefix version. It should be noted that the ending "-ose" implies a free anomeric center, which is not present in CDs. A systematic nomenclature was proposed where cyclic oligosaccharides composed of a single type of residue could be named "*by giving the systematic name of the glycosyl residue, preceded by the linkage type in parentheses, preceded in turn by 'cyclo-' with a multiplicative suffix*" (e.g., cyclohexakis-(1,4)-α-D-glycosyl for α-CD) [1].

Large CDs have been given Greek letters as prefix by French and co-workers, as a natural continuation of the generic names of the α-, β- and γ-CDs. However, the Greek alphabet is finite and will not be able to accommodate the growing number of large CDs described. The last CD to benefit from a Greek letter prefix will be ω-CD (cyclomaltononacosaose, CD_{29}) (Table 1.1). Moreover, although researchers are familiar with the generic names for the small CDs, α-, β- and γ-CDs, the use of the Greek letter prefix for the large CDs is confusing and non-descriptive of the size of the macrocycle. Large CDs have often been designated "cycloamylose" (abbreviated CAn, where n designated the number of glucose molecules in the macrocycle). However, this is a non-systematic name whose use has been discouraged. The designation "LR-CD" has often been used to distinguish large CDs from large derivatives of α-, β- and γ-CDs [1].

1.2.2. Classification

Apart from the naturally occurring CDs, to extend their usefulness in several applications, many branched CDs and CD derivatives have been synthesized by chemical or enzymatic modification to change the physical and chemical properties of the CDs, modifying their solubility, complex-forming capacity, thermal properties and chemical stability [3, 8].

In CDs, every glucopyranose unit has three free OH groups, two of which are secondary (C-2 and C-3) and one primary (C-6). As each of these free hydroxyl groups can be modified, by substituting the hydrogen atom or the hydroxyl group by a wide variety of substituents, the majority of simple synthetic reactions results in a considerable number of positional isomers [6, 8].

Branched CDs (or second-generation CDs) can be obtained by chemical synthesis but in most cases they are prepared by enzymatic reactions. Branched CDs can be divided into two categories: homogeneous and heterogeneous branched CDs, or single- and multiple-branched CDs (Fig. 1.1). Homogeneous branched CDs have only glucose or malto-oligosaccharide side chains bound to the native CDs. Heterogeneous branched CDs have one or more galactose or mannose residues bonded either to each other or directly to the parent CD rings. The

Table 1.1.   Nomenclature and some properties of CDs [1, 7].

Glycosyl units	Semisystematic name	Generic name	Abbreviation	Molecular weight	Aqueous[a] solubility [g/100 mL]	Surface[a] tension (mN/m)	Specific rotation $[\alpha]_D^{25}$	Half life of[b] ring opening (h)	Radius of[c] gyration (Å)
6	cyclomaltohexaose	α-cyclodextrin	α-CD	972.9	14.5	72	+147.8	33	6.0
7	cyclomaltoheptaose	β-cyclodextrin	β-CD	1135.0	1.85	73	+161.1	29	6.7
8	cyclomaltooctaose	γ-cyclodextrin	γ-CD	1297.2	23.2	73	+175.9	15	7.3
9	cyclomaltononaose	δ-cyclodextrin	CD9	1459.3	8.19	73	+187.5	4.2	—
10	cyclomaltodecaose	ε-cyclodextrin	CD10	1621.4	2.82	72	+204.9	3.2	—
11	cyclomaltoundecaose	ζ-cyclodextrin	CD11	1783.6	>150	72	+200.8	3.4	—
12	cyclomaltododecaose	η-cyclodextrin	CD12	1945.7	>150	72	+197.3	3.7	—
13	cyclomaltotridecaose	θ-cyclodextrin	CD13	2107.9	>150	72	+198.1	3.7	—
14	cyclomaltotetradecaose	ι-cyclodextrin	CD14	2270.0	2.30	73	+199.7	3.6	—
15	cyclomaltopentadecaose	κ-cyclodextrin	CD15	2432.2	>120	73	+203.9	2.9	—
16	cyclomaltohexadecaose	λ-cyclodextrin	CD16	2594.3	>120	73	+204.2	2.5	—
17	cyclomaltoheptadecaose	μ-cyclodextrin	CD17	2756.4	>120	72	+201.0	2.5	—
18	cyclomaltooctadecaose	ν-cyclodextrin	CD18	2918.6	>100	73	+204.0	3.0	—
19	cyclomaltononadecaose	ξ-cyclodextrin	CD19	3080.7	>100	73	+201.0	3.4	—
20	cyclomaltoeicosaose	o-cyclodextrin	CD20	3242.9	>100	73	+199.7	3.4	—
21	cyclomaltoheneicosaose	π-cyclodextrin	CD21	3405.0	>100	73	+205.3	3.2	11.5

(Continued)

Table 1.1.   (*Continued*)

Glycosyl units	Semisystematic name	Generic name	Abbreviation	Molecular weight	Aqueous[a] solubility [g/100 mL]	Surface[a] tension (mN/m)	Specific rotation $[\alpha]_D^{25}$	Half life of[b] ring opening (h)	Radius of[c] gyration (Å)
22	cyclomaltodoicosaose	ρ-cyclodextrin	CD22	3567.2	—	—	—	—	—
23	cyclomaltotriicosaose	σ-cyclodextrin	CD23	3729.3	—	—	—	—	—
24	cyclomaltotetraicosaose	τ-cyclodextrin	CD24	3891.4	—	—	—	—	—
25	cyclomaltopentaicosaose	υ-cyclodextrin	CD25	4053.6	—	—	—	—	—
26	cyclomaltohexaicosaose	φ-cyclodextrin	CD26	4215.7	—	—	—	—	19.6
27	cyclomaltoheptaicosaose	χ-cyclodextrin	CD27	4377.9	—	—	—	—	—
28	cyclomaltooctaicosaose	ψ-cyclodextrin	CD28	4540.0	—	—	—	—	—
29	cyclomaltononaicosaose	ω-cyclodextrin	CD29	4702.2	—	—	—	—	—
30	cyclomaltotriacontaose	—	CD30	4864.3	—	—	—	—	—
31	cyclomaltohentriacontaose	—	CD31	5026.5	—	—	—	—	—
n	—	—	CD*n*	*n*·162.14	—	—	—	—	—

[a] Observed at 25°C.
[b] In 1 mol/L HCl at 50°C.
[c] Determined by small angle X-ray scattering at 25°C.

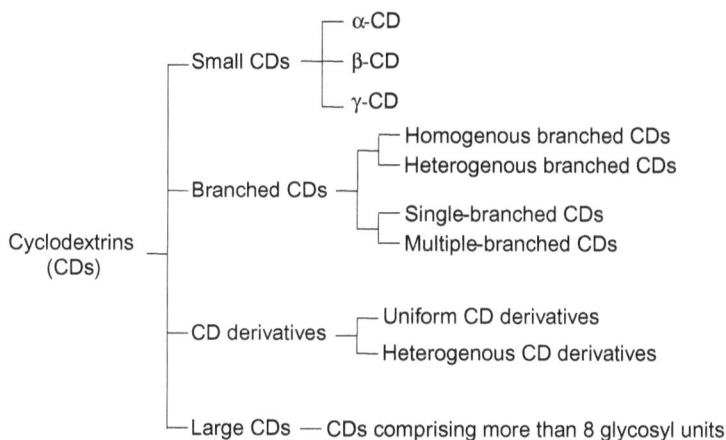

Fig. 1.1.   Classification of CDs.

solubilities of branched CDs in water, even in aqueous 80% ethanol or in aqueous 50% solutions of methanol, formaldehyde and ethylene glycol, are extremely high in comparison with their parent CDs [8].

The derivatives usually are produced by aminations, esterifications or etherifications of primary and secondary hydroxyl groups of the CDs. Depending on the substituent, the solubility of the CD derivatives is usually different from that of its parent CDs. Virtually all derivatives have a changed hydrophobic cavity volume and also these modifications can improve solubility, stability against light or oxygen and help to control the chemical activity of *guest* molecules [2].

Modification of the CDs either with glucosyl, maltosyl, hydroxypropyl, hydroxyethyl, methyl or sulfate groups increases their aqueous solubility. Modification of CDs with low aqueous solubility or those that are insoluble in water can be achieved by adding aliphatic groups or short nonpolar groups to the cyclodextrin or by cross-linking CDs with a suitable cross-linker, such as epichlorohydrin, to form spherical beads of polymers. These modified CDs have the same functional properties as the unmodified CDs such as stabilization of *guests*, etc. [3].

Some inorganic esters of CDs, such as nitrates, sulfates, phosphates etc. have also been synthesized; however, the organic (acetyl, benzoyl, propionyl, methyl and carbamoyl) esters have been more frequently used for practical industrial and analytical purposes [6, 8]. Ether derivatives are the most important CD derivatives from a practical point of view. These compounds can be prepared either by direct reaction with an alkylating agent or via an intermediate, such as sulfonate esters or deoxy-halogeno derivatives. This group equally contains neutral, anionic and cationic derivatives as well as silyl ethers [8].

Deoxy CDs can be classified according to the mode of preparation into two groups: intermediaries which can be further derivatized, and end products containing thio-, amino-, substituted amino- or azido-substituents [8]. Hydroxyl protons can be exchanged by deuterium at the oxygen–hydrogen bonds or carbon–hydrogen bonds in deuterated CD derivatives. These molecules represent the smallest group of CD derivatives and have found application only in NMR studies of CDs [6, 8].

1.2.3. *Structure*

The three major CDs (α-, β- and γ-CDs) are crystalline, homogeneous, and non-hygroscopic substances, which are torus-like macro-rings built up from 6, 7 and 8 glucopyranose units, respectively (Fig. 1.2). The ring that constitutes the CDs, in reality, is a cylinder, or better said a conical cylinder, which is frequently characterized as a doughnut or wreath-shaped truncated cone [4, 9].

The cavity is lined by the hydrogen atoms and the glycosidic oxygen bridges. The non-bonding electron pairs of the glycosidic oxygen bridges are directed toward the inside of the cavity producing a high electron density there and lending to it some Lewis base characteristics [9]. From the X-ray structures, it appears that

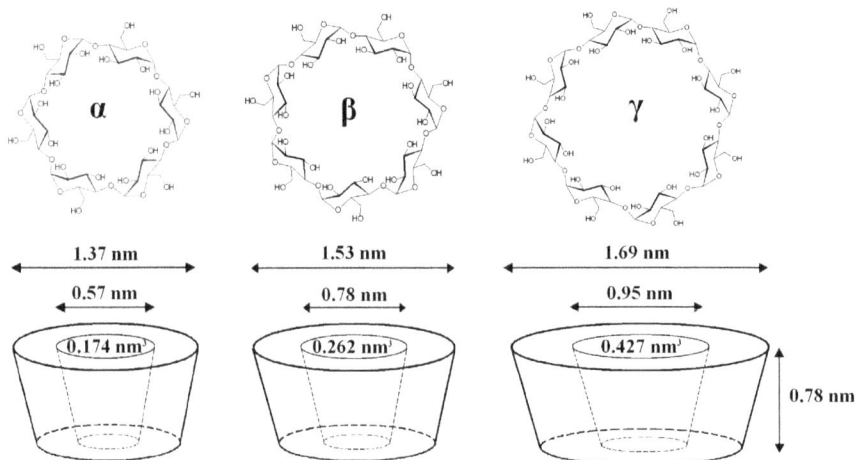

1.37 nm	1.53 nm	1.69 nm
0.57 nm	0.78 nm	0.95 nm
0.174 nm³	0.262 nm³	0.427 nm³

0.78 nm

Fig. 1.2.   Structures, approximate geometric dimensions and approximate cavity volumes of α-, β- and γ-CD [1, 9].

in CDs the secondary hydroxyl groups (C-2 and C-3) are located on the wider edge of the ring and the primary hydroxyl groups (C-6) on the other edge, and that the apolar C-3 and C-5 hydrogens and ether-like oxygens are at the inside of the torus-like molecules. This results in a molecule with a hydrophilic outside, which can dissolve in water, and an apolar cavity, which provides a hydrophobic matrix, described as a "micro heterogeneous environment" [2].

The C–2–OH group of one glucopyranose unit can form a hydrogen bond with the C–3–OH group of the adjacent glucopyranose unit. In the CD molecule, a complete secondary belt is formed by these H-bonds; therefore the β-CD has a rather rigid structure. This is probably the explanation for the observation that β-CD has the lowest water solubility of all CDs. The H-bond belt is incomplete in the α-CD molecule, because one glucopyranose unit is in a distorted position. Consequently, instead of the six possible H-bonds, only four can be established fully. The γ-CD is a non-coplanar, more flexible structure; therefore, it is the most soluble of the three CDs [9]. On the side where the secondary hydroxyl groups are situated, the diameter of the cavity is larger than on the side with the primary hydroxyls, since free rotation of the latter reduces the effective diameter of the cavity (Fig. 1.2).

Nuclear magnetic resonance (NMR), infrared (IR) and optical rotary dispersion (ORD) spectroscopy studies have demonstrated that D-glucopyranose units have the same conformation in both dimethyl sulfoxide (DMSO) and heavy water (D$_2$O). The spectroscopic studies on CDs in aqueous solution suggest that the conformation of CDs in the solution is almost identical to their conformation in the crystalline state. β-CDs have perfect symmetry, while α- and γ-CDs rings are slightly distorted. The planar structure of α-CDs in solution deviates by about 100 pm from that in the crystal structure. The twist form observed experimentally in the glucose unit number five out of the plane of the five other glucose units disappears in solution [4, 9].

The solid state structures of CD$_9$, CD$_{10}$, CD$_{14}$ and CD$_{26}$ have been reported. The structure of CD$_9$ exhibits a distorted elliptical boat-like shape, but it retains a similar structure to regular CDs. CD$_{10}$ and CD$_{14}$ also exhibit a more elliptical macrocyclic ring folded in a saddle-like shape. The structure of CD$_{26}$ has channel-like cavities composed of two short V-amylose helices in anti-parallel orientation, and its structure is very different from the regular CDs [1, 7] (Fig. 1.3).

In contrast to the crystal structures of the small CDs, which are often regarded as representative frozen images of their conformation in solution, the structures of CD$_9$, CD$_{10}$, CD$_{14}$ and CD$_{26}$ as found in their crystalline forms can only partly be taken as representative structures. Owing to the large flexibility of these molecules, a

Large cyclodextrin   Top-view   Side-view

CD9

CD14

CD26

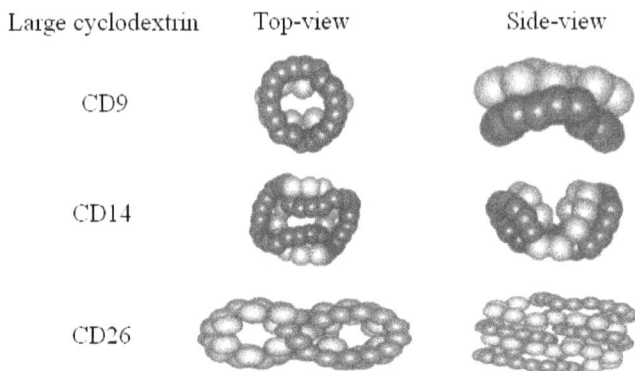

Fig. 1.3.   Structures of some large CDs.

very wide range of structural conformations, including band-flips, can be expected to occur in solution [1].

Studies on the conformation of CD_{21} in the solution by small-angle X-ray scattering (SAXS) with molecular modeling simulations showed that this molecule as most likely adopted a circularized three-turn single helical structure with a radius of gyration of 11.5 Å [1]. Other LR-CDs structures have not been reported, because their single crystals could not be prepared. However, several LR-CDs have been deduced from molecular dynamics simulations and SAXS analysis [1, 7].

1.2.4. *Property*

The water solubility of CDs is unusual. β-CD is at least nine times less soluble than α- and γ-CDs (Table 1.1). The thermodynamic properties of α- and γ-CDs are similar. The decreased solubility of β-CD in water appears to be due to the marked structure of water arising from water–β-CD interactions, causing a compensation of the favorable enthalpy by the unfavorable entropy of solution [9].

The solubility of CDs depends strongly on the temperature. As the temperature increases, the aqueous solubility of the CDs increases. Mathematically, this is shown in Eq. (1) for α-CDs, Eq. (2) for β-CDs and Eq. (3) for γ-CDs, where c is the concentration of CD in mg/mL and T is the temperature in K [8, 9].

$$c_{\alpha\text{-CD}} = (112.71 \pm 0.45)e^{-(3530\pm31)[(1/T)-(1/298.1)]} \tag{1}$$

$$c_{\beta\text{-CD}} = (18.3236 \pm 0.099)e^{-(14137\pm31)[(1/T)-(1/298.1)]} \tag{2}$$

$$c_{\gamma\text{-CD}} = (219.4 \pm 9.8)e^{-(3187\pm320)[(1/T)-(1/298.1)]} \tag{3}$$

The CD_9 had greater aqueous solubility than β-CD, but less than that of α- and γ-CDs. It was the least stable among the CDs known at this time; their hydrolysis rate increases in the order of α-CD $<\beta$-CD $<\gamma$-CD $<CD_9$ [4, 9]. With the exceptions of CD_9, CD_{10} and CD_{14}, the aqueous solubility of large CDs is very large compared to their linear counterparts. The solubilities of CD_9, CD_{10} and CD_{14} are intermediary to those found for α- and β-CD. While the solubility of CD_{11} to CD_{13} and CD_{15} to CD_{17} exceeds 150 g and 120 g per 100 mL, respectively, the solubility of CD_{14} and CD_{10} is very low, compared to that of β-CD. Both regular CDs and LR-CDs show no surface activity [1].

β-CD is not soluble in most organic solvents, but is soluble in some solvent/water mixtures. In general, for most solvents, the solubility of the CD decreases as the concentration of non-aqueous solvent increases. The exceptions are ethanol, propanol and acetonitrile where a maximum solubility occurs at a 20% to 30% concentration of the organic solvent [3].

Although CDs are more resistant to acid hydrolysis than starch, strong acids, such as hydrochloric acid and sulfuric acid hydrolyze CDs yield a series or mixture of oligosaccharides ranging from an opened ring down to glucose. The rate of hydrolysis increases as the concentration of acid and temperature increase. Hydrolysis is minimal or below the limits of detection in the presence of weak acids such as organic acids [3]. The acid catalyzed hydrolysis rate for α- through CD_{17} indicates that the stability of the macrocyclic ring decreases with increasing number of glycosyl units, probably owing to increased flexibility and a higher number of decomposition points (α-D-1,4 linkages) [1].

CDs are not hydrolyzed by bases, even at elevated temperature. Exposure to a 0.35 mol/L sodium hydroxide solution at 60°C results in nil detectable hydrolysis of CD [3]. CDs will undergo oxidative reactions. The periodate reaction opens the glucose ring but no formaldehyde or formic acid is formed since no reducing groups are present. β-CD is rapidly and completely oxidized at 50°C by hypochlorite at 5.25% of bleach solution. β-CD is oxidized less rapidly by hydrogen peroxide. Only 11% of the CD is oxidized after 2 h at 50°C when CD is exposed to 30% of hydrogen peroxide. Exposure of CD to 50–200 ppm of hydrogen peroxide used to prevent microbial growth results in no detectable oxidation of the CD [3].

α- and β-CDs are more resistant to hydrolysis by α-amylase than starch. CD glucosyltransferase can hydrolyze as well as form CDs. Enzymes such as β-amylases and glucoamylases requiring a reducing end group cannot hydrolyze CDs. Many α-amylases will not attack α- and β-CDs, and those which do, will hydrolyze the CD more slowly than starch. Unlike other CDs, γ-CD is readily hydrolyzed by α-amylases. CDs are biodegradable. Many microorganisms found in soil can degrade CDs [3]. Owing to their inherent flexibility, the large CDs are readily susceptible to enzymatic degradation by amylolytic enzymes [1]. Interestingly, the half-lives of ring opening for CD_9 to CD_{17} parallel their $^{13}C1$ and $^{13}C4$ chemical shifts [1].

1.3. Inclusion Complex Formation, Preparation and Characterization

1.3.1. *Formation of inclusion complexes*

CDs can be considered as empty capsules of a certain molecular size. Figure 1.2 illustrates the approximate volumes of each capsule (α-, β-, γ-CDs). As a result of this, CDs are able to form "*inclusion complex*" with a wide variety of hydrophobic *guest* molecules, which is the key to their successful applications. One of the molecules, the "*host*", includes, totally or partly, the "*guest*" molecules by physical forces (Fig. 1.4) [2, 9].

Complexation is a molecular phenomenon where one molecule of *guest* and one molecule of CD come into contact with each other to associate and form a complex [3, 10]. The beneficial effects of complexation of a *guest* with CD include increased solubility of the *guest*; stabilization of the *guest* to prevent volatilization, oxidation, and degradation due to exposure to light and heat; elimination or

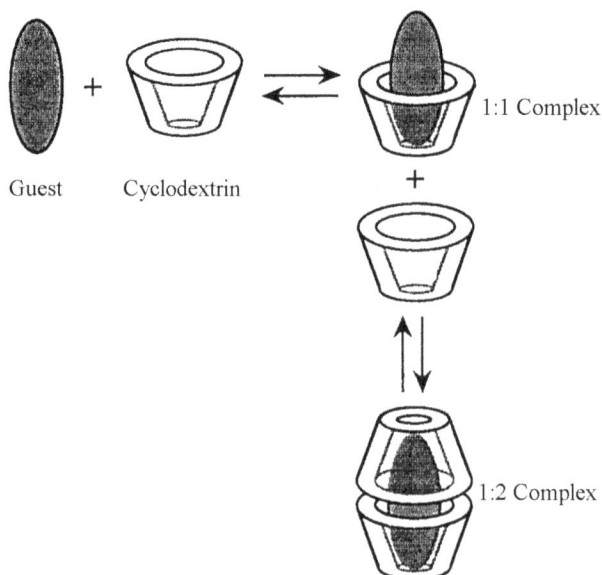

Fig. 1.4. Scheme illustrating equilibrium binding of *guest* and CD to form a 1:1 complex, and a 1:1 complex with a second molecule of CD to form a 1:2 complex [10].

reduction of undesired tastes or odors; prevention of chemical reaction; directed chemical synthesis and separation and isolation of various chemicals [3].

The inclusion of a *guest* in a CD cavity consists basically of a substitution of the included water molecules by the less polar *guest*. The process is energetically favored by the interactions of the *guest* molecule with the solvated hydrophobic cavity of the *host*. In this process entropy and enthalpy changes have an important role [9]. In spite of the fact that the "driving force" of complexation is not yet completely understood, it seems that it is the result of various effects [9]:

(a) Substitution of the energetically unfavored polar–apolar interactions (between the included water and the CD cavity on one hand, and between water and the *guest* on the other) by the more favored apolar–apolar interaction (between the *guest* and the cavity) and the polar–polar interaction (between bulk water and the released cavity-water molecules).
(b) CD-ring strain release on complexation.
(c) van der Waals interactions and hydrogen bonds between *host* and *guest*.

The type of bond established between the *guest* and the *host* is non-covalent. A variety of non-covalent forces such as van der Waals forces, hydrogen bonding, dipole–dipole interaction, London dispersion forces and other hydrophobic interactions are responsible for the formation of a stable complex. Also, a force responsible for complexation for one series of molecules may not hold for another series of molecules. A single force cannot be found as the only factor responsible for complexation of all molecules. Several forces seem to be involved in the complexation. As a result, it is difficult or impossible to predict how well a particular molecule might bind with CDs [2, 3, 9].

Looking for the most appropriate *host*, CDs with various cavity diameters and chemically modified CDs must be discriminated. Using the higher soluble CD derivatives, high solubilizing effects can be attained [9].

The geometric compatibility between *host* cavity and *guest*, the structure, charge and polarity of the *guest*, the effect of the reaction medium (solvent) and temperature are important for inclusion complex formation [9]. It is important that the geometrical dimensions of the *guest* molecules are rather close to those of substituted benzene ring or its condensed homologues [9]. The larger the *guest* molecule, the slower is the formation and decomposition of the complex. The charge and polarity of the *guest* also play an important role in the CD-substrate *host–guest* interaction; however, it is less decisive than that of the geometric fitting. Ionization decreases the rate of complex formation and decomposition. This recombination–dissociation equilibrium is one of the most important characteristics of this association [9]. Molecules can be complexed by CDs when they are less hydrophilic or less polar than water, and there is a positive correlation between stability of CD complexes and the hydrophilic character of molecules or certain parts of the *guest* molecules. In the case of the charge, the complexation of neutral molecules is easier than the ionized counterpart [9]. The role of the medium in which complex formation takes place plays an important role, since its presence influences the inclusion equilibrium process in both directions. The CD complex formation does not require any extra solvents, but the presence of at least a minimum amount of water is necessary for the inclusion processes [9].

In contrast to the numerous studies on the small CDs, the complex forming properties of the vast range of large CDs are still to a great extent unknown. The larger CDs are not regular cylinder shaped structures and their rings have to be highly flexible. They are collapsed, and their real cavity is even smaller than in the γ-CD. As the CD cavity diameter increases, this nonpolar cavity can accommodate an increasing number of water molecules, and in the aqueous solution these "complexed" water molecules will differ energetically less and less from the bulk of the solvent. As a consequence, complex formation in such a system does not result in a significant gain in energy. The driving force of the complex formation, the substitution of the high enthalpy water molecules in the CD cavity, is weaker in the case of larger CDs [4, 9].

The inclusion complex formation constants between LR-CD (CD_9–CD_{17}) and various anions have been measured by capillary electrophoresis. The findings showed that LR-CDs have a certain extent of inclusion ability [1, 7]. β-CD was the best overall complex former with the chosen range of *guest* molecules followed by α- and γ-CDs. The stability constants decreased for CD9 and CD10. CD10 revealed itself as the poorest overall complex former with the *guest* molecules studied (e.g., 4-*tert*.-butyl benzoate and the ibuprofen anion) [1].

The inclusion complex forming ability of the large CDs varies greatly according to their size, suggesting that they, just like α-, β- and γ-CDs, are able to present more or less suitable cavities depending on the size and structure of the guest molecules. For example, the increased stability constants observed for the complexes formed between 4-*tert*.-butyl benzoate and the ibuprofen anion and CD_{11} to CD_{17}, indicates that these molecules are able to present a more suitable cavity for small guest molecules, compared to CD_9 and CD_{10}.

Somehow, the large CDs must be able to present a pseudo-cavity of similar size, in order to form stable inclusion complexes. The introduction of more glycosyl units may increase the strain of the molecule due to torsion angle limitations. This increased strain may cause the formation of a weaker complex, as indicated by the experimental data. A study of the complex formation between even larger CDs (CD_{18}–CD_{23}) and these or similar *guest* molecules may confirm this hypothesis, as it may be expected that the 1:1 stability constants will display a local maximum around CD_{20}–CD_{21} [1].

Kitamura and co-workers studied the complex formation of CD_{21} to CD_{31} with triiodide (I_3^-) using isothermal titration calorimetry. A more elaborate model assuming 1:2 complex formation with identical interacting sites was employed instead of 1:1 complex formation. The data suggested that this range of large CDs is able to accommodate two triiodide molecules [1, 7].

The complex formation between a LR-CD mixture with a DP from 22 to more than 100 and iodide and other chemical compounds has been reported. LR-CD mixtures with a DP from 22 to 45 and over 50, respectively, exhibited an efficient artificial chaperone for protein refolding. The findings showed that chemically denatured citrate synthase was folded within 2 h, and over 50% of its activity was recovered within 30 min. As a consequence, a protein refolding kit using a LR-CD mixture came onto the Japanese market. This product was the first practical application of LR-CDs. To investigate further applications of LR-CDs, it is necessary to prepare large amounts of each isolated pure LR-CD efficiently [1, 7].

1.3.2. *Methods of preparation of inclusion complexes*

Water is important in the formation of complexes. In addition to being a driving force for the hydrophobic interaction of the *guest* with the cavity of the CD, water is a medium for the dissolution of both the CD and the *guest* [11]. In some cases, water is required to maintain the integrity of the complex. Water is present in the crystals of the complex. The water can form a bridge between the hydroxyl groups of adjacent molecules of CD to form a cage to assist in trapping the *guest* in the complex [11].

Complexes are easily formed. The most commonly used methods are co-precipitation, slurry, paste and dry mixing methods. They are all similar, with each method using successively less water [11].

The co-precipitation method is the most widely used method in the laboratory. A solution of CD is made, and the *guest* compound is added to the solution while stirring. Conditions are selected so that the solubility of the complex is exceeded and the complex can then be collected as a precipitate by filtration or centrifugation. For less soluble CD, the solution is heated to dissolve the CD before adding the *guest*. The CD solution is allowed to cool to ambient temperature as it is stirred with the *guest* to form the complex. The collected complex can be dried [11].

The CD need not be completely dissolved to form complexes. For the slurry method, CD is suspended in water up to 40%–45% w/w concentration. The *guest* can have an effect upon the viscosity of the slurry, and concentrations are adjusted to allow mixing of the CD and *guest*. Heating can be used if desired and is compatible with the *guest*. The time is dependent upon the particular *guest* and the vigorousness of the stirring. The complex is generally collected by filtration and dried if a dry complex is required [11].

The paste method uses a minimum amount of water, 20–30% w/w, which is about the same amount of water as is found in the filter cake from a complex made by the co-precipitation method. The CD, water and *guest* are added to a mixing device and mixed. Because of the high viscosity, this method is usually not performed in the laboratory. The mixing time varies with the *guest* to be complexed, amount of water, and mixing device used. In most cases, a mixing time of about 30 min is sufficient, depending upon the mixer selected and the *guest*. The complex is then dried without any further treatment. In some cases, this method can be done in two steps — a mixing step followed by a holding step to allow completion of the complexation reaction. In some literatures, this is referred to as the heat and seal method, because the *guest* and CD mixture is placed in a container and sealed [11].

The dry mixing method involves mixing the CD with the *guest* with no added water. This is generally not an efficient method of making complexes since mixing times can range from hours to days. There are some exceptions, such as lemon oil, where complexation is completed in a few minutes. In these cases, the *guest* might also be serving as a solvent for the CD [11].

When drying complexes, it is desirable to remove the water as quickly as possible, especially when a volatile *guest* has been complexed. There is no single method or process for making complexes. A process must be developed for each *guest* to be complexed with the CD. Selection of equipment is frequently based upon the use of existing available equipment, and a process is then developed around the use of that equipment. Factors such as temperature, amount of water, mixing time and drying conditions must be established for the equipment and each *guest* and CD used to optimize the process [11, 12].

1.3.3. *Characterization of inclusion complexes*

A variety of techniques are used to analyze complexes. Only NMR proves that a complex is formed. A shift in peaks can be observed for both the CD and the *guest*. As the environment around the hydrogen atoms in the cavity changes with association with the *guest*, a shift in the peaks for the CD can be observed. Similarly, shifts can be observed for peaks corresponding to the atoms of the *guest* which penetrate into the cavity of the CD [11].

Most users of complexes do not have access to an NMR and use other means to characterize the complex. While these methods do not prove that a complex has been formed, evidence can be obtained that is consistent with a complex having been formed [11].

Extraction of the *guest* from the complex is frequently used to determine the load and the homogeneity of the complex. A small quantity of complex is placed into a container with some water and a water-immiscible solvent and heated and mixed thoroughly. The heat destabilizes the complex. The CD becomes solubilized in the water, and the *guest* is extracted into the organic phase. The *guest* in the organic phase is assayed using the chromatographic or spectrophotometric procedures normally used for the assay of the *guest* [11].

Differential scanning calorimetry (DSC) or thermogravimetric analysis (TGA) can also be used. For analysis by these techniques, the *guest* must have a melting or boiling temperature below about 300°C, the temperature at which the CDs decompose. Using DSC, no energy absorption is observed at the melting temperature of the *guest* when the *guest* is complexed. Since the *guest* is surrounded by the CD and not interacting with other *guest* molecules, there is no crystalline *guest* structure to absorb energy. In both these techniques, an increase in the boiling temperature is observed. Interaction of the *guest* with the CD provides a higher energy barrier to overcome volatilization so that an increase in boiling temperature of about 10°C is observed. An estimation of the amount of noncomplexed *guest* can be obtained from DSC, especially if large amounts of *guest* are not complexed [11, 12].

Fourier transform infrared spectroscopy (FTIR) and Raman spectroscopy have also been used for analyses of complexes. Upon complexation of the *guest*, shifts or changes in the spectrum occur. There are interferences in the spectra from the CD, and some of the changes are very subtle, requiring careful interpretation of the spectrum [11].

Several other techniques have also been used to characterize complexes, such as ORD, circular dichroism, mass spectroscopy, fluorescence and X-ray crystallography. Selection of some of these specialized techniques frequently depends on equipment available and properties of the *guest* which make the particular technique most sensitive or reliable for the particular complex [11].

Some functional assays can also be used. For labile *guests*, accelerated stability tests can be done to determine if decomposition of the *guest* is prevented or is within the expected range for a complex. Since most *guests* complexed by CDs are hydrophobic, their interaction with water is altered so that tests measuring wet ability, dissolution rate or even rate of release of *guest* from some complexes can be used [11, 12].

References

 1. Larsen, KL (2002). Large cyclodextrins. *Journal of Inclusion Phenomena and Macrocyclic Chemistry*, 43, 1–13.
 2. Martin Del Valle, EM (2004). Cyclodextrins and their uses: A review. *Process Biochemistry*, 39, 1033–1046.
 3. Shieh, WJ and AR Hedges (1996). Properties and applications of cyclodextrins. *Journal of Macromolecular Science, Part A: Pure and Applied Chemistry*, 33(5), 673–683.
 4. Szejtli, J (1998). Introduction and general overview of cyclodextrin chemistry. *Chemical Review*, 98, 1743–1753.
 5. Hashimoto, H (2002). Present status of industrial application of cyclodextrins in Japan. *Journal of Inclusion Phenomena and Macrocyclic Chemistry*, 44, 57–62.
 6. Szejtli, J (2004). Past, present, and future of cyclodextrin research. *Pure and Applied Chemistry*, 76(10), 1825–1845.
 7. Ueda, H (2002). Physicochemical properties and complex formation abilities of large-ring cyclodextrins. *Journal of Inclusion Phenomena and Macrocyclic Chemistry*, 44, 53–56.
 8. Astray, G, C Gonzalez-Barreiro, JC Mejuto, R Rial-Otero and J Simal-Gandara (2009). A review on the use of cyclodextrins in foods. *Food Hydrocolloids*, 23, 1631–1640.
 9. Cserhati, T and E Forgacs (2003). *Cyclodextrins in Chromatography*, pp. 1–10. Cambridge: The Royal Society of Chemistry.
10. Rajewski, RA and VJ Stella (1996). Pharmaceutical applications of cyclodextrins. 2. In vivo drug delivery. *Journal of Pharmaceutical Sciences*, 85(11), 1142–1169.
11. Hedges, AR (1998). Industrial applications of cyclodextrins. *Chemical Review*, 98, 2035–2044.
12. Giordano, F, C Novak and JR Moyano (2001). Thermal analysis of cyclodextrins and their inclusion compounds. *Thermochimica Acta*, 380, 123–151.

2
ENZYMES IN PREPARING CYCLODEXTRINS

Sheng-Jun Wu*, Xiu-Ting Hu†, Jin-Moon Kim‡ and Jing Chen§

**School of Food engineering, HuaiHai Institute of Technology
59 Cangwu Road, Lianyungang 222005, China
† The State Key Laboratory of Food Science and Technology,
School of Food Science and Technology, Jiangnan University,
Wuxi 214122, China
‡Food Safety and Inspection Services, Office of Field Operations
US Department of Agriculture, 230 Washington Avenue Extension
Albany, NY 12203, USA
§ Yihai-Keny Institute of Food Technology, Hebei Qinhuangdao economic
and Technological Development Zone (Eastern)
1 Jinhai Road 066206, China*

2.1. Introduction

Enzymatic process is the main method to produce cyclodextrins (CDs), and so far, chemical methods have been reported. CD is industrially produced from starch, glucogen, malto-oligosaccharides, and other dextrins through catalysis by cyclodextrin glucosyl transferase (CGTase). More and more researchers focus on the essential CGTase to prepare CD due to its wide applications in food, medicine, cosmetics, environmental protection, and analytical chemistry.

2.2. CGTase

By subjecting starch to *Bacillus macerans*, F. Schardinger obtained CD in 1903, and he found that this process was carried out in the presence of an enzyme, i.e., CGTase. In 1939, Tilem and Hudson confirmed this enzyme, and named CGTase (E.C.2.4.1.19) or cyclo-starch transferase. It was also named *cyclic Bacillusmacerans* amylase in some reports due to being earlier found in *cyclic B. macerans*. Tilben *et al.* prepared CGTase from *B. macerans* in 1942. As the investigation on CD continued, more CGTase producing strains were isolated from nature. *Gram-positive bacteria* are responsible for producing CGTase, while minority of *Gram-negative bacteria* are capable of synthesizing high activity CGTase. In 1960s, Horikoshi obtained β-CD through alkaline fermentation without solvent, and the yields reached 75%–80%. Thanks to this result, the cost of CD production decreased dramatically in the early 1980s. Based on the study on α-CD producing CGTase, Kabayashi Ensuiko Sugar Refining Corporation Ltd (Japan) developed a continuous production process which increased CD yields and achieved industrial CD production. These two findings greatly promoted the CD industry [1].

2.2.1. *The catalytic mechanism of CGTase*

2.2.1.1. *The structure of CGTase*

2.2.1.1.1. The primary structure of CGTase

The catalytic mechanism of CGTase is closely related to the structure of its protein. Experiments showed that its nucleophilic attack sites focused on the Asp residue, and that the catalytic proton donor was Glu residue. van der Veen *et al.* found that CGTase showed rather explicit feature of its primary structure, and different sources of CGTase had 47%–99% similarity in the amino acid sequence [2]. Machovic and Janecek found that there were 51 conserved sequences of amino acid residues located in the five conserved amino acid sequence regions of α-amylase family through comparison of the amino acid sequence of 31 CGTases. The arrangement of conserved region I is shown in Table 2.1. These residues are closely related to

Table 2.1.   The arrangement of conserved region I of α-amylase family [3].

Enzymes	Strain source	The arrangement of conserved region I
CGTase	*Bacillus circulans* strain *251*	VIIDFAPNH
α-amylase	*Aspergillus oryzae*	LMVDVVANH
Starch invertase	*Neisseria polysaccharea*	LMMDLVVNH
Branching enzyme	*Escherichia coli*	VILDWVPGN
isoamylase	*Pseudomonas amyloderamosa*	VYMDVVYNH
Maltose amylase	*Thermus* strain IM6501	VMLDAVFNH
Oligomeric-1,6- glucosidase	*Bacillus cereus*	LMMDLVVNH

the catalytic activity, substrate binding capacity and calcium ion binding capacity of CGTase [3].

The differences in primary structure of varying sources of CGTase are closely related to the product-specificity, but the mechanism has not yet been elucidated [3].

2.2.1.1.2. The domain and active centers of CGTase

Starch-degrading enzyme can be divided into α-amylase family GH13, β-amylase family GH14 and glucoamylase family GH15 according to primary structure and structural domain. With the development of investigations, a number of new enzymes have emerged. For example, pullulanase, isoamylase, neopullulanase and other members were added to α-amylase family GH13. The amylase family GH13, GH70 and GH77 can be classified as glycoside hydrolase family GH-H, and CGTase (EC2.4.1.19) was member of GH-H, which hydrolyzed starch to produce CD. Among all GH-H family members, there are four common characteristics: (a) Catalytic structural domain possesses $(\beta/\alpha)_8$ barrel structure, which is also called TIM-barrel structure, and is often referred to as structure domain A, and is a eight parallel β-sheet in closed state outsourcing eight α-helix; (b) Similarly conserved regions, including the consistency of the secondary structure, especially the conformity of β-sheet; (c) The fourth β-sheet of aspartic acid functions as a nucleophile, and the fifth β-sheet of the glutamic acid functions as a proton donor, and the seventh β-sheet of the aspartic acid maintains the stability of substrate binding when the enzymes perform a catalytic role; (d) The same mechanism of breaking of α-glycosidic bond.

As a member of the starch hydrolase family, CGTase also has these characteristics, but the enzymes from different sources have their own special features. However, a number of different structural domains of CGTase are different from each other. For example, CGTase from *B.circulans*, *Klebsiella oxytoca* and *Pyrococcus furiosus* has 5, 4 and 4 structural domains, respectively, while most of amylases only have two AB structural domains, i.e., α-amylase structural domain. A typical CGTase domain possesses five structural domains (A–E structural domain): AB is the catalytic structural domain, among which A possesses $(\beta/\alpha)_8$ barrel structure, while B is embedded between the third β-strands and the third α-helix of domain, and the amylase active center is composed of AB at the C-terminal of β-sheet layer of AB; C possesses substrate binding and stability functions, and has a similar structure of the C-terminal with that of amylase structure; E has two maltose binding sites (MBS), and both Trp-616 and Trp-662 are relatively important in MBS1 which has large capacity of binding amylose, while Tyr-633 is relatively important in MBS2 that is responsible for guiding the substrate to the active center.

Several tryptophans (Trp-590, Trp-615) in the whole structural domain played relatively important roles in the hydrolysis of starch, and they are responsible for guiding amylose to the enzyme's active center. For a while, there was no progress in gaining knowledge in the function of D. In 2005, through the domain-shuffling methods, Rimphanitchayakit *et al.* investigated the function of each structural domain by assembling all structural domains of CGTase from different strains, and found that the function of D was to guide E to AB, determine the space orientation of E and AB thus affecting the enzyme reaction [4]. For example, as to *B. macerans,* Davies *et al.* found that, in the activity center of CGTase, nine substrate binding sites ranged in order in the substrate binding groove and numbered +2 to −7 [5]. Uitdehaag *et al.* confirmed that the enzyme catalytic sites were at the substrate binding site +1 to −2 elaborating the cyclization mechanism of CGTase. When the catalytic reaction takes place, the non-reducing end of substrate binds to the substrate binding sites −7, and the substrate binding sites +1, also the activator binding sites, activates glycosyltransferase reaction of the enzyme through an induced-fit mechanism [6]. The structure of CGTase is shown in Fig. 2.1.

2.2.1.2. *The catalytic mechanism of CGTase*

CGTase is a kind of glycoside hydrolases, and according to the deference in the conformation of the anomeric carbon atom of glycosidic bonds subjected to hydrolysis, there are two kinds of enzyme catalytic mechanisms: One is conformation inversion type (Fig. 2.2(a)), and the other is conformation keeping

Fig. 2.1.   The structure of CGTase [3].

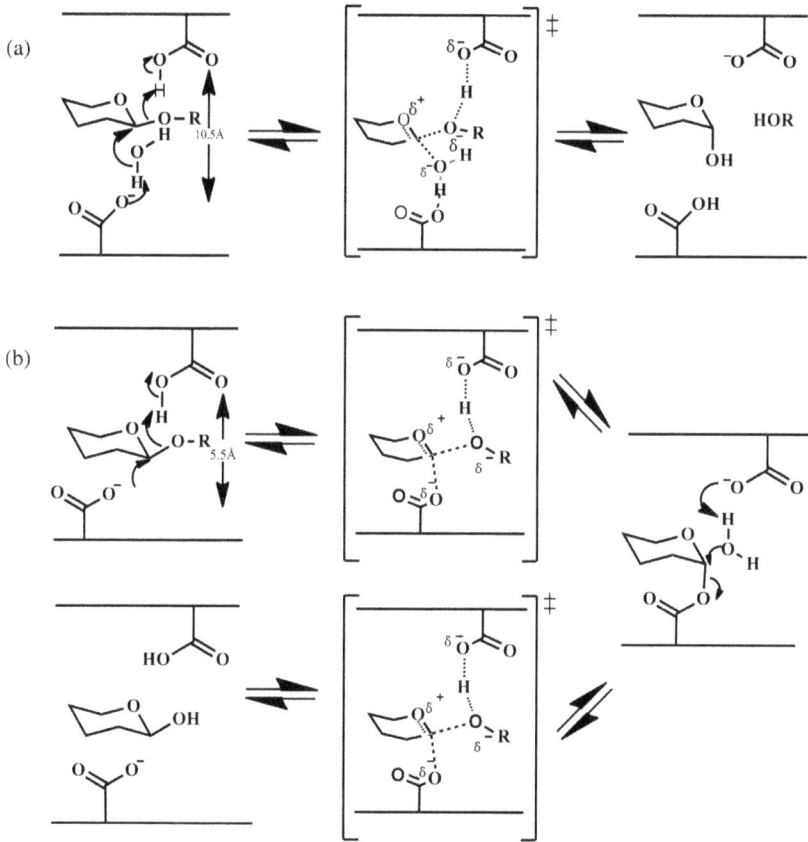

Fig. 2.2. The cleavage modes of glycosidic bond catalyzed by CGTase [6].

type (Fig. 2.2(b)). With conformation inversion type, water molecules attack the anomeric carbon atom as nucleophile during the reaction: An amino acid of the enzyme protein, usually glutamic acid, as proton donor, results in the transformation of anomeric carbon atom (from β- to α-) during the opening of the glycosidic bond, and the reaction is not reversible. As to conformation keeping type, another amino acid in the enzyme protein (usually aspartic acid) attacks the anomeric carbon atom as nucleophile, and a covalent bond will be formed between it and the substrate. Thus, an intermediate of the enzyme and substrate is formed, and then water molecules attack the intermediates as nucleophilic, in which one of the hydrogen atoms combines with amino acids and a covalent bond is formed between the hydroxyl and the anomeric carbon atom proton. Thus, the hydrolysis of the substrate is finished while the conformation of the anomeric carbon atom in the product remains unchanged. In the case of the hydrolysis reverse reaction of conformation keeping type, it becomes a glycosyltransferase reaction mechanism if the water molecule is replaced by another glycosylation, and this is why as a member of α-amylase family but CGTase has catalytic reactivity of glycosyltransfer reaction.

Glycosylation transfer reaction catalyzed by CGTase is divided into four types: hydrolysis reaction, cyclization reaction, coupling reaction and disproportionation reaction, in which the coupling reaction in some documents is also known as the open-loop reaction. The direction of glycosylation reaction catalyzed by the CGTase depends on the reaction rate constant of the four types of reaction. Typically, the rate constant of cyclization reaction is greater than that of the hydrolysis reaction, and therefore the direction of the reaction is mainly to produce CD (Fig. 2.3).

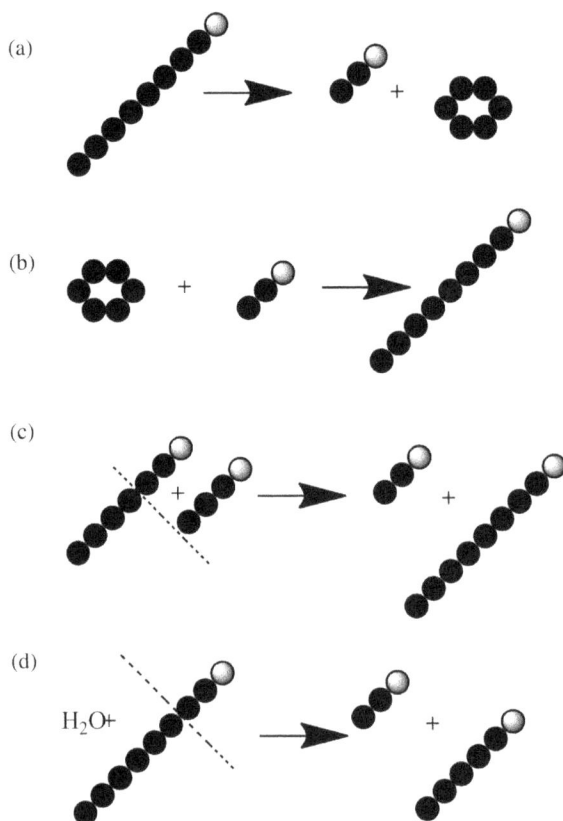

Fig. 2.3. The schemes of the four kinds of reaction [6]. a: Cyclization reaction; b: Coupling reaction; c: Disproportionation reaction; d: Hydrolysis reaction.

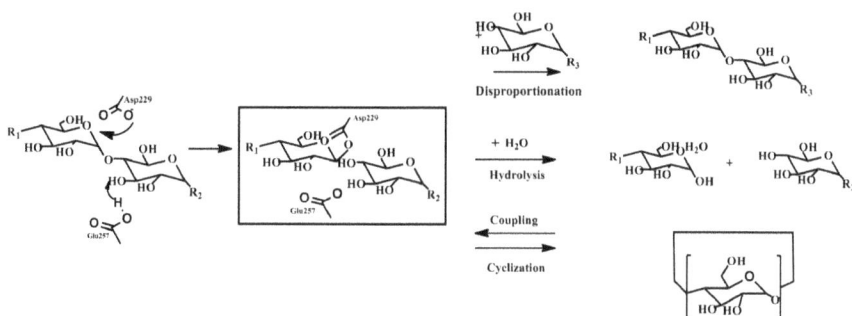

Fig. 2.4. The mechanism of the four kinds of reaction catalyzed by CGTase [6].

In Fig. 2.4, R_1, R_2, R_3 represent different chemical groups, i.e., a number of glucosyls connected by α-1, 4 glycosidic linkages, where n represents the number of sugar residues ($n \geq 6$). Cyclization reaction is a unique reaction catalyzed by CGTase, i.e., to form the CD by linking the oligosaccharide chain, enzymatically released from the substrate, end-to-end. The reverse reaction of cyclization is coupling reaction, and can also be catalyzed by such enzymes. In practice, a substrate with long chains can promote the cyclization reaction, while high concentrations of malto-oligosaccharides or glucose as substrate can promote the coupling reaction. If natural starch is used as the substrate, the initial reaction is mainly disproportionation, leading to rapid decrease of the viscosity of the substrate solution; when appropriate length of the chain of substrate appears, cyclization will once again become a dominant reaction. On the other hand, as a member of α-amylase family, CGTase also possesses some of the starch hydrolysis activity, and can catalyze the hydrolysis of starch. With respect to the cyclization process of CGTase, the latest view is that these enzymes randomly cleave α-D-$(1, 4)$ glycosidic bond in amylose, and link the non-reducing end with the reducing end of newly formed oligosaccharide chain. Thus, in the initial reaction, CDs with glucose residues ranging from 6 to 60 are formed, but in the subsequent reaction, common α-CD, β-CD and γ-CD are formed from the larger molecular weighted CDs through the cyclization reaction and catalyzed by CGTase. The mechanism of cyclization reaction of CGTase is the di-substitution reaction, in which α-glycosidic bond is retained, and can be divided into five steps:

a. When the substrate is combined with the enzyme activity center, Glu257 provides a proton to the glucose residue at the substrate binding site $+1$, and the proton is combined with the oxygen atom of glycosidic bond of glucose residues, on the other hand, Asp229 residue reacts with the C1 of non-reducing end glucose residue at substrate binding site -1;

b. A transition-state material in the form of oxygenated phosphorus cation forms, followed by the formation of covalent intermediate;

c. At the substrate binding site +1, the glucose residue combined with hydrogen protons comes off the enzyme active center, leading to the break of the substrate strand, and then the newly formed reducing-end-glucose residue attacks the covalent bond between non-reducing-end-glucose residue and the Asp229 residue;

d. The transition-state material in the form of oxygenated phosphorus cation forms again;

e. Reducing-end-glucose residue provides a hydrogen proton to the Glu257 residue, and takes the place of the Asp229 residue, and is then combined with the non-reducing-end-glucose residue to form a new α-D-(1,4)-glycosidic bond and some hydroxy, resulting in cyclization.

During the reaction, a group of the side chain of Glu257 functions as a general acid catalyst to provide hydrogen protons; nucleophilic groups Asp229 residues stabilize intermediate and Asp328 plays an important role in combining with the substrate.

2.2.1.3. *The advances in the product specificity of CGTase research*

So far, the nature of the genes of CGTase still has some unresolved questions. First, the general view on the bacteria which produce CGTase is to transform the surrounding carbon material into CDs which the normal microorganisms cannot utilize, while these bacteria can often synthesize the proteins that can transport and utilize the CDs, for their survival with advantages over other bacteria. However, in some bacterial species, only CGTase genes were found, and the corresponding genes encoded proteins that can transport and utilize the CDs were not observed. In this regard, the meaning of CGTase secretion to the bacterium itself has not yet been explained, and the possible causes may be gene deletion or special use of CD during the evolution of bacteria, e.g., cellular signal transduction. Second, during the whole-genome sequencing of some bacterial species that can produce and make use of CD, it is found that CGTase gene may not exist in the bacterial chromosome, e.g., CGTase genes of CGTase-producing *Xanthomonas axonopodis pv. citri* strain were not found during whole-genome sequencing. The reasonable explanation of the exact location and high expression mechanism of CGTase genes of such bacteria from an evolution point of view has yet to be studied further.

In the GM study of CGTase, the research on evolution of product-specificity of the enzyme is the most mentioned. Those amino acid residues for substrate binding in the CGTase active site are considered to have an important influence on the product specificity of these enzymes, and are often chosen as the evolution of experimental mutation, and the location of these amino acid residues can be determined by analysis of the spatial structure of these enzymes combined with competitive inhibitor. Among all amino acid residues for binding with substrate, a part of all kinds of CGTase and even the α-amylase family are highly conserved, and is considered to have little effect on the product specificity of these enzymes. Mutation selection focuses on the region where some amino acid residues have relatively poor conservation, such as substrate binding region sites −7, in which all types of CGTases differ significantly in primary structure. The area, most of the amino acid residues used for substrate binding is in the amino acid ring at the beginning of functional domain B, i.e., between His140 residue (the highly conserved residue in the α-amylase family) and Glu153 residue (the residue of CGTase specific). In order to disrupt the substrate, a combination of the amino acid residues in this region, Ser146Pro of CGTase of *B. circulans* 251 strains, caused β-dextrin production to decrease in proportion, while the ratio of α-CD increased by this enzyme. In this region, the number of amino acids residues of γ-CGTase is less by six than that of α- and β-CGTase. According to this nature, amino acid residues 145–151 of CGTase produced by *B. circulans* strain 8 were replaced. Thus, it was found that the ability of the enzyme to produce γ-CD had increased. The mutation is considered to disrupt the ability of the enzyme to transform the initially high molecular CD into small one, thus contributing to synthesis of greater CD molecules.

Another amino acid residue area with a relatively poor conservation is the region of substrate binding site −3, as mentioned above. Various types of CGTases in this region are significantly different in primary structure. Analysis of the spatial structure of the enzyme after combination with inhibitors indicated that the Tyr89 residue in the ring of amino acid residues 87–93 in this region is able to combine with the substrate through hydrophobic reaction. It was found that through mutation of Tyr89Asp Ill of CGTase from the strains of *B. circulans* 251, the ratio of α-CD production catalyzed by this enzyme increased slightly. Mutation of Tyr89Phe of CGTase from *Bacillus sp.* I-5 strain increased the ratio of β-CD production, but led to a decline in the overall output, while the mutation of Tyr89Ser of the same enzyme has no effect on product specificity. The Asn94 residue in this area differs in all kinds of CGTase, and is considered to have influence on the product specificity of the enzyme. Mutation of Asn94Ser of α-CGTase from *Bacillus sp.* I-5 strain improved the ratio of α-CD production. Thus, the overall production of CD increased. Double mutation of Ser146Pro/Tyr89Asp at the substrate binding site 3 and −7 of CGTase from *B. circulans* 251 strain increased the ability of the enzyme to produce α-CD in one-fold.

Currently, despite some research progress in the evolution of product specificity of CGTase, there are still many questions to be answered. First, most of the transformation led to the reduction in product-specificity of enzymes, and cannot transform the product specificity of the enzyme reasonably and directly, and more understanding on the spatial structure and the function of amino acids before and after the transformation of the enzyme is needed to solve this problem. Second, experiments on the evolution of product specificity have a negative impact on the cyclization activity of these enzymes perhaps to some extent, because mutations mostly occur near the active ring center in the present experiments, and therefore, the new idea of experimental design of full account of the nature of many regions may help solve this problem. Thirdly, experimental mutations of the same or corresponding amino acid residues often lead to quite different results, indicating that the functions of the specific amino acid residues in the enzyme may include many aspects. All such aspects of the nature of the mutated amino acid residues should be considered when designing mutation experiments, and the study may also provide a new perspective to the determinants of product specificity of the enzyme. Fourthly, previous research in this field mainly employed traditional means of enzyme evolution and experimental design perspectives, and means were relatively single and points of view of experimental design were of some limitations, and therefore new methods and ideas need to be established. In 2005, Rimphanitchayakit *et al.* built a multi-enzymes chimeric gene through combination of gene of *β*- CGTase from *B. circulans* strains and that from *Paenibacillus macerans* IAM1243 strains in different functional domains, and the structure and product specificity were analyzed. In these experiments, the chimeric enzymes showed different activity, and part of the product specificity of chimeric of enzymes changed. The results showed that the region of amino acids having important effect on the product specificity of the enzyme was in the central domains of A and B, counting from the C-terminal. Experiments also suggested that the functional domains C and D could be close to functional domains A and B, and correctly made spatial orientation of the functional domain E, resulting in an important effect on the activity of the enzyme. This research has brought a new understanding of the determinants of the nature and product specificity of CGTase, and also suggested the important functions of domains C and D [4].

2.2.2. *Classification of CGTase and their bacteria sources*

2.2.2.1. *Classification of CGTase*

In the whole enzyme system, according to the catalyst type of enzyme, the known enzymes are divided into six categories:

a) Oxidoreductases

$$RH + R'(O_2) \rightarrow R + R'H(H_2O)$$

b) Transferases

$$RG + R' \rightarrow R + R'G$$

c) Hydrolase

$$RR' + H_2O \rightarrow RH + R'OH$$

d) Lyases

$$RR' \rightarrow R + R'$$

e) Isomerases

$$R \rightarrow R'$$

f) Ligases or Synthetases

$$R + R' + ATP \rightarrow RR' + ADP(AMP) + P_i(PP_i)$$

This classification system was proposed by the Enzyme Commission of the International Union of Biochemistry in 1955, and is now generally accepted by the academic world. In this classification system, according to the specific type of reaction or the nature of substrates, each category is further divided into sub-categories and sub-sub-categories, and a four-digit coding system is used. Each enzyme has a four-digit code, in which the first digit indicates categories, the second digit represents sub-class of, the third digit means sub-sub-categories and fourth-digit states the number of sub-sub-categories. The enzymatic number of CGTase is E.C.2.4.1.19, and the taxonomy name is 1,4-α-D-glucopyranosyltransferase. The enzyme is a sugar synthase with multiple catalytic sites, and according to its catalytic mechanism, it belongs to both glycosyltransferase and α-amylase families. This is mainly because CGTase catalyzes glycosyl transfer reactions, both intramolecular and intermolecular, but also has α-amylase activity in nature, and hydrolyzes the straight-chain glucan connected by α-1, 4 - glucosidic linkages.

According to the number of glucose residues, the CDs can be divided into α-CD, β-CD and γ-CD, etc, and accordingly, CGTases are mainly divided into α-, β- and γ-CGTase based on the types of CDs produced by enzyme catalysis. These three types of CGTases have wide applications in the industry. However, regardless of the types of CGTase used, a CD mixture will be formed at the final equilibrium due to non-exclusive product specificity. Types of CDs produced in the early fermentation in which CGTase acts on the substrate can be determined by the type of enzyme, i.e., one of α-, β- and γ-CGTase. This classification of

CGTase is rather rough, and the nature of CGTase from varied source is different. For example, the preparation process and α-CD yield can be different depending on the source of α- CGTase from *B. macerans* or *Bacillus stearothermophilus* in the preparation of α-CD. Therefore, on the classification of CGTase, the source of the bacteria must be taken into consideration.

2.2.2.2. *Bacteria source of CGTase*

Since the first discovery of CGTase secreted by *B. maceran*, CGTase has been successively isolated from a variety of microorganisms, such as *B. macerans, B. stearothermophilius, Bacillus megaterium, Bacillus licheniformis, B. circulams, Bacillus subtilis, Bacillus firmus lentus, Klebsiella Pp neumoniae, Micrococcus spp., Pseudomonas spp.* and *Alcalibacterium spp.*

In industrial production, a lot of bacteria can be used for the production of CGTase, including aerobic mesophilic bacteria such as *B. circulars* and *B. megaterium*, etc, aerobic thermophilic bacteria such as *B. stearothermophilus*, and anaerobic thermophilic bacteria, such as *Thermoanaerobacterium thermosulfurigenes*, aerobic alkaliphilic bacteria such as *B. circulars, Bacillus fat*, and aerobic halophilic bacteria such as *halophilic bacilli*, etc. The CGTase secreted by most of the microbes are extracellular enzymes, and the yield of the CD and the main product of these enzymes are different, mainly α-CD, β-CD and γ-CD.

2.2.3. *The enzymatic properties of CGTase*

CGTase from different sources of bacteria exhibit different properties. The molecular weight, optimum pH, optimum temperature, isoelectric point and the main product of CGTase are summarized in Table 2.3. The data are of instructively scientific significance for the production of CD. For example, the molecular weight of CGTase produced by strains of *B. circulans* IFO 332 reaches a maximum of 200,000 Da, while that of CGTase from strains of *B. circulans* C3110 is the smallest (only 30,000 Da), indicating that there is difference in the number of amino acid residues and degree of polymerization of the enzyme. Optimum pH of CGTase from *B. circulans* C3110, *K. pneumoniae, B. circulans* DF9R and *B. macerans*, etc, is in the acidic range. That of CGTase from *B. subtilis* no.313, *Alkalipic B. sp.* ATCC21783, *Brevibacterium sp.* no.9605 and *Bacillus agaradhaerns* LS-3C, etc, is in the alkaline range. In contrast, CGTase from *Bacillus lentus* and *Alkalipic B. sp.* ATCC8514 show optimum pH in the neutral range. The differences in optimum temperature of CGTase are large, and the range of the optimum temperature of CGTase reported in the literatures is even broader.

Fujita *et al.* found that the CGTase produced by the *Bacillus. sp.* AL-6 remained stable at pH 5.0–8.0 and temperature 40°C for 3 h [7], and Norman and Joergensen also found that the CGTase from *Thermoanaerobacter spp.* can exist stably at 100°C in the presence of starch [8]. The isoelectric points for CGTase from varied strains show large differences. The largest difference reaches 8.8 and the smallest is only 2.8. The main products of various CGTases are different. However, for more kinds of CGTases, the main products are α- and β-CDs. For less CGTases, the main product is γ-CD, relatively. In industrial production, these properties are undoubtedly beneficial, and we can select the appropriate enzyme depending on the target product.

2.2.3.1. *The nature of CGTase from Bacillus alkalophilus 1177*

2.2.3.1.1. Molecular weight

After subjected to purifying, the CGTase secreted by *B. alkalophilus* 1177 and the standard proteins were subjected to SDS-PAGE electrophoresis simultaneously.

According to the SDS-PAGE electrophoresis performed on standard proteins with known molecular weight (Fig. 2.5), standard protein relative mobility was plotted against logarithm molecular weight of standard protein as shown in Fig. 2.6.

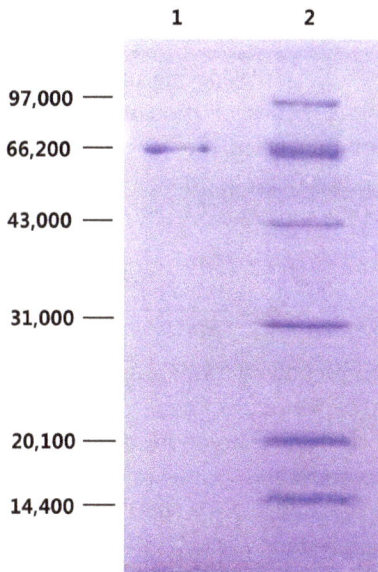

Fig. 2.5. The plot of the SDS-PAGE electrophoresis of pure CGTase from *B. alkalophilus* and standard proteins [7].

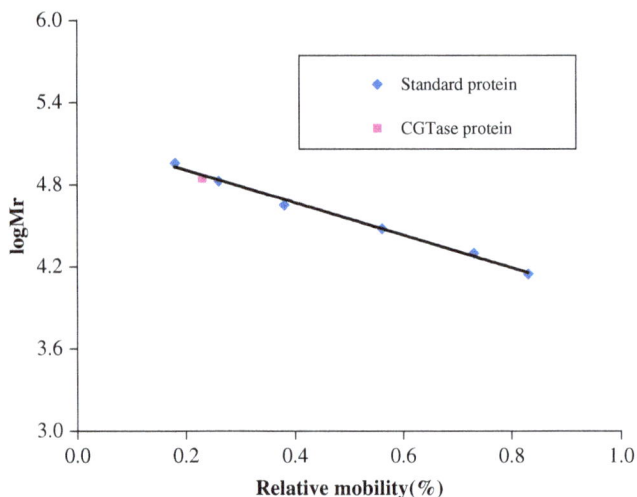

Fig. 2.6. Determination of the molecular weight of the CGTase from *B. alkalophilus* [7].

Then, regression equation ($\log Mr = -1.1977 R_f + 5.1504$) was obtained ($R_f$ is the migration rate). The molecular weight of the enzyme was calculated to be 69 kD.

2.2.3.1.2. Kinetic constants (Km and Vmax)

Km is a constant characteristic of the enzyme, only with the nature of the enzyme, and has nothing to do with the concentration of the enzyme. Its physical meaning is the substrate concentration when the enzyme reaction rate reaches half of the maximum reaction rate. Michaelis–Menten equation quantitatively describes the relationship between the substrate concentration and the maximum reaction rate.

$$V = \frac{V \max \cdot [S]}{Km + [S]},$$

where V is the enzymatic reaction rate, Vmax is the maximum enzymatic reaction rate, and $[S]$ is the substrate concentration.

By using the reciprocal of the concentration of starch and the reciprocal of the reaction rate as the abscissa axis and the vertical axis, respectively, we obtained the double reciprocal curve: $y = 0.0108x + 0.0087$, $R^2 = 0.9996$ (Fig. 2.7), where x is the reciprocal of the concentration of starch substrate, and y is the reciprocal of the reaction rate. According to the double reciprocal plot method of Linewear–Burk (i.e., draw a straight line and extrapolated to intersect with the abscissa axis, and intercept in the abscissa axis was $-1/Km$, and the intercept of in vertical axis was

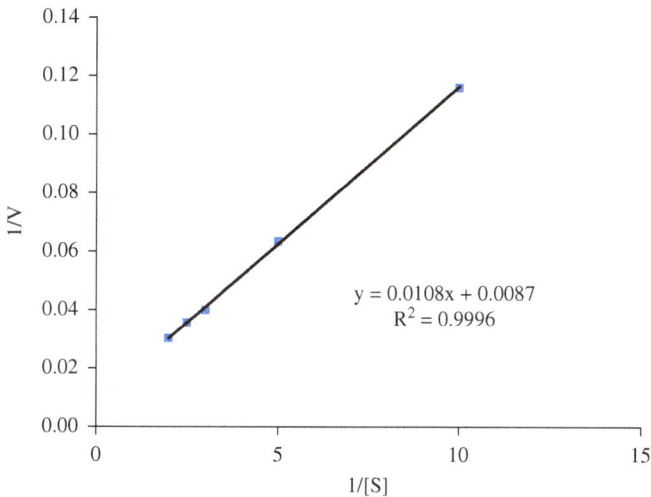

Fig. 2.7. The double reciprocal curve of CGTase [7].

$1/V$max), we obtained the Km of 1.24 mg/mL and Vmax of 114.9 μg/min using starch as substrate of enzyme. In other previous reports, the Km of the enzyme produced by *B. firmus* NCIM 5119, *B. circulans* E192 and *B. macerans* ATCC 8514 was 1.21 mg/mL, 5.7 mg/mL and 3.33 mg/mL, respectively. The Km measured in this study was relatively smaller, indicating that this enzyme has a strong affinity for starch.

Langmuir mapping method in recent years has been recommended, and is more evenly distributed in the plot. This method is based on S/V plot against S (i.e., draw a straight line and extrapolate to intersect with the abscissa axis, $-Km$ and Km/Vmax are the intercepts of abscissa axis and vertical axis, respectively and $1/V$max is the slope) (Fig. 2.8). According to this curve, the Km and Vmax were obtained, 1.09 mg/mL and 103.1 μg/min, respectively, which were less than those derived from Linewear–Burk double reciprocal plot method.

2.2.3.1.3. Optimum pH and pH stability

The enzyme activity was measured by the conventional method after the enzyme solution was diluted with different pH buffers (acetate buffer 4–5, phosphate buffer 6–8, glycine–NaOH buffer 8.5–11). The residue activity was again measured in the same way after the enzyme solution was incubated at different pH buffers and 40°C for 30 min.

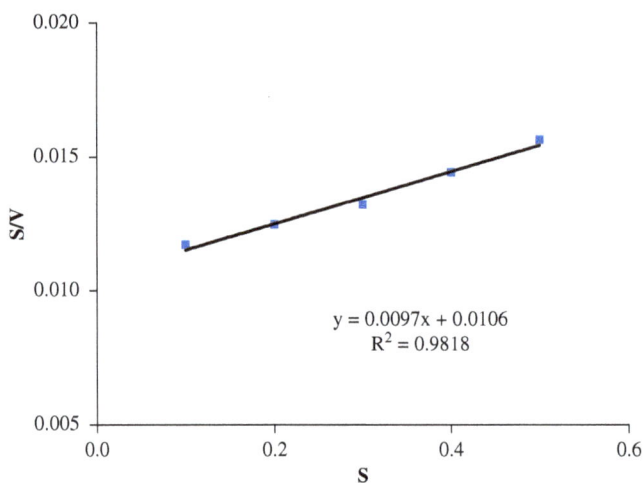

Fig. 2.8.   The Langmuir curve of CGTase [7].

$$y = 0.0097x + 0.0106$$
$$R^2 = 0.9818$$

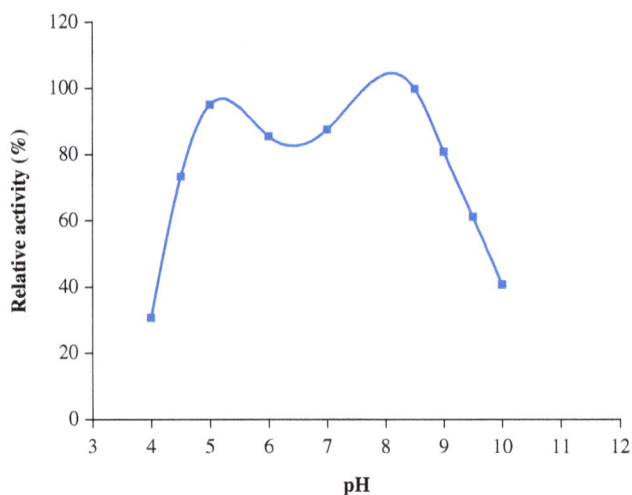

Fig. 2.9.   Effect of pH on CGTase activity [10].

As shown in Fig. 2.9, the relative activities of were 97% and 100%, respectively, at the two peaks: pH 5.0 and pH 8.5. In contrast, optimum pHs of the enzyme were reported at about 5 and 9 [9], and 4.5 and 7.0 [10]. It can be seen from Fig. 2.10 that the enzyme was basically stable in the range of pH 6.0–10.0, and maintained 90% of activity.

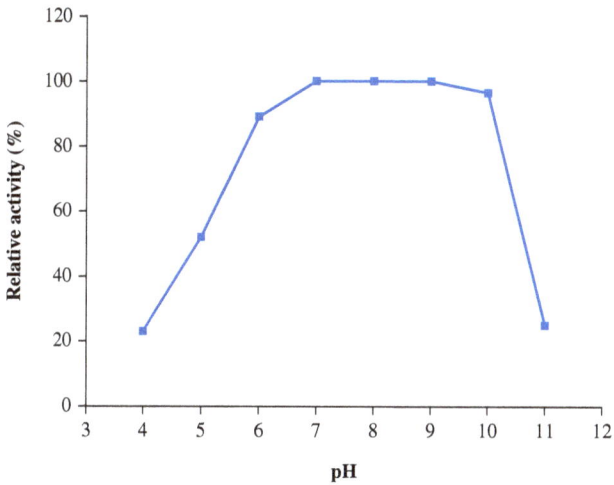

Fig. 2.10. pH stability of CGTase [10].

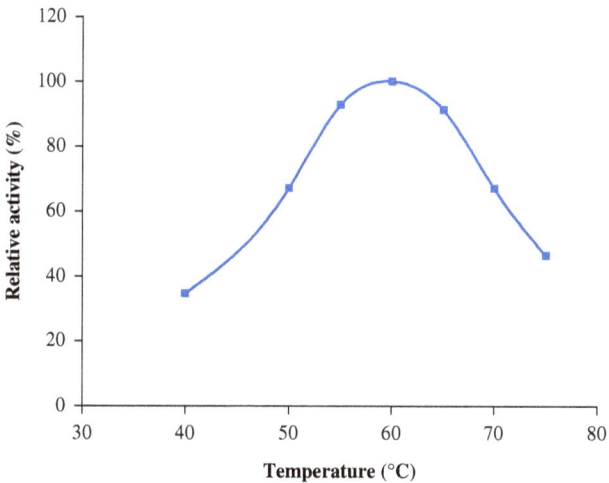

Fig. 2.11. Effect of temperature on CGTase activity [10].

2.2.3.1.4. Optimum temperature and thermal stability

The enzyme activity measured at different temperatures at 30°C, 40°C, 50°C, 60°C, 70°C and 80°C, respectively, to observe the optimum temperature of enzyme, and the thermal stability of the enzyme was determined by incubation (pH 8.5) at different temperatures for 30 min. As shown in Fig. 2.11, the optimum temperature

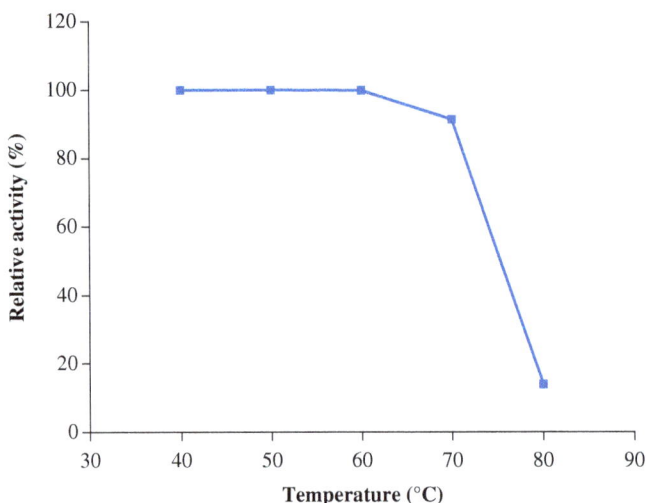

Fig. 2.12. Thermal stability of CGTase [10].

for the enzyme was 60°C. The enzyme remained stable below 70°C, and began to lose activity above 70°C (Fig. 2.12).

2.2.3.1.5. Effect of metal ions on the enzyme activity

The enzyme activity was significantly inhibited by Ag^+, Cu^{2+}, Mg_2^+, Al^{3+}, Co^{2+}, Zn^{2+}, Fe^{2+}, etc, and inhibited by Sn^{2+}, Mn^{2+} to a certain extent, while other metal ions did not affect the enzyme activity (Table 2.2).

2.2.3.2. *The nature of CGTase from B. alkalophilus sp.G1*

2.2.3.2.1. Molecular weight

As shown in Fig. 2.13, the SDS-PAGE electrophoresis of CGTase from *B. alkalophilus sp.* G1 showed a single protein band, and by comparing with the standard protein bands, the molecular weight of CGTase was judged to be 75 kDa.

2.2.3.2.2. Kinetic constants (*Km* and *V*max)

After purification, 0.1 mL of the CGTase solution from the *B. alkalophilus sp.* G1 was incubated in 1 mL of 0.1 M phosphate buffer (pH 6.0) (the concentration of soluble starch ranged from 0.4–6.0 mg/mL) at 60°C for 10 min, and the kinetic diagram of CGTase based on the Hanes equation is shown in Fig. 2.14. *Km* and

Table 2.2. Effect of metal ions on CGTase activity [10].

Metal ion	Final concentration (mmol/L)	Relative enzyme activity (%)
Ag^+	1.0	27
Cu^{2+}	1.0	35
Sn^{2+}	1.0	78
Mn^{2+}	1.0	83
Mg^{2+}	1.0	46
Ca^{2+}	1.0	98
K^+	1.0	97
Al^{3+}	1.0	31
Co^{2+}	1.0	37
Zn^{2+}	1.0	34
Fe^{2+}	1.0	36

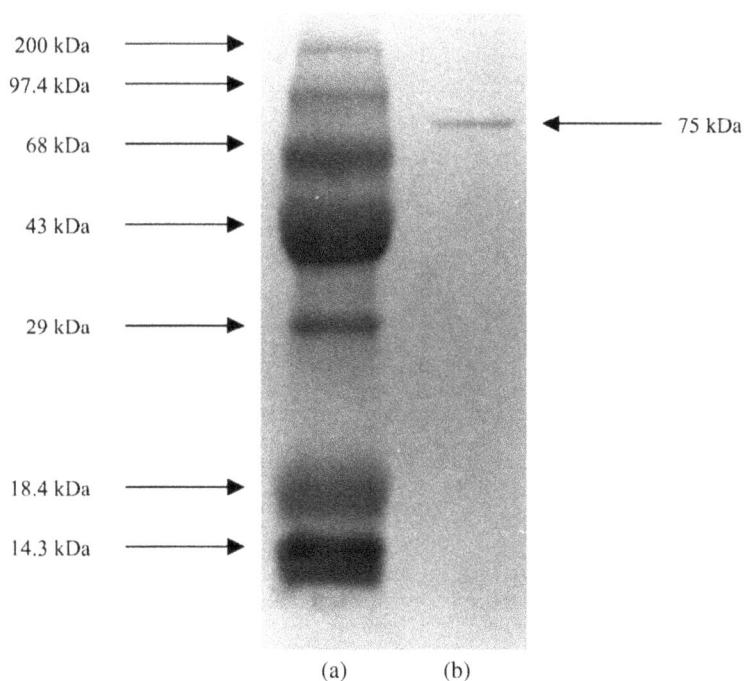

Fig. 2.13. The SDS-PAGE electrophoresis of pure CGTase standard protein [10].

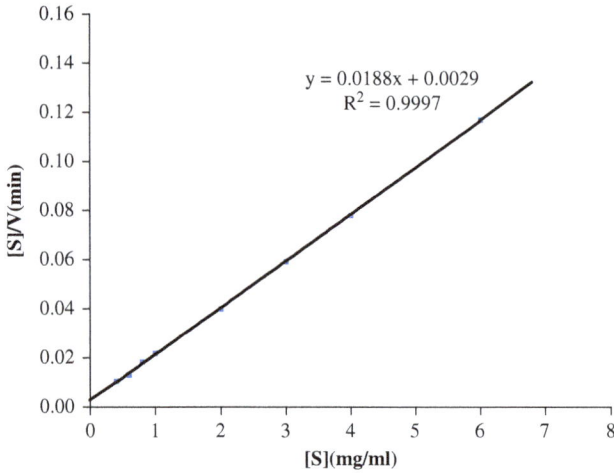

Fig. 2.14. Pure CGTase kinetic diagram (according to Hanes equation) [10].

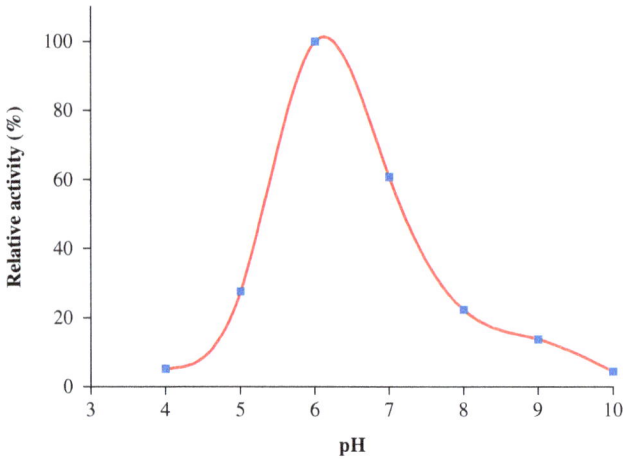

Fig. 2.15. Effect of pH on CGTase activity [10].

Vmax were calculated to be 0.15 mg/mL and 60.39 mg β-CD/mL/min according to the equation.

2.2.3.2.3. Optimum pH and pH stability

As shown in Fig. 2.15, according to classical analysis method of the enzyme activity of the CGTase, the optimum pH of the enzyme is found to be 6.0. Figure 2.16

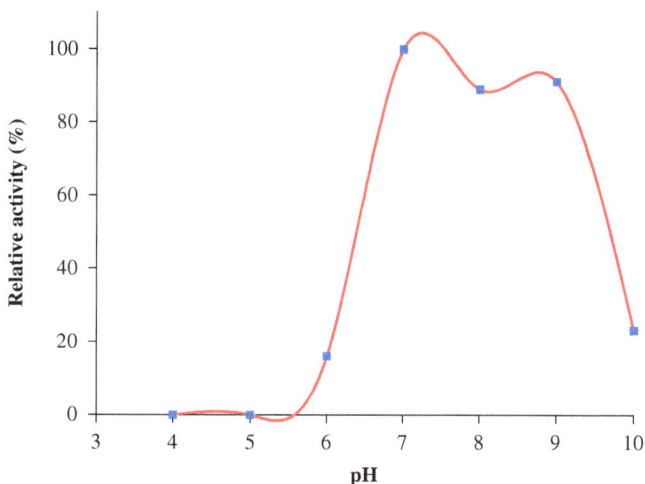

Fig. 2.16.   pH stability of CGTase [10].

also shows that the enzyme is active only in the neutral pH range, and the enzyme ring activity will be lost at extreme pH conditions. Most of the optimum pHs of the CGTase currently reported are concentrated in the range of 5.0–8.0, of course, there are some bacteria, such as the optimum pH of the CGTase from *basophilis* can reach above 10.0 [10].

In the case of determination of the pH stability of the enzyme, a series of 0.1 mL of purified enzyme solution were added to 0.2 mL of the three types of buffer with different pHs, i.e., 0.1 M sodium acetate (pH 4.0–5.0), 0.1 M sodium phosphate (pH 6.0–7.0) and amino acid and sodium hydroxide buffer (pH 9.0–10.0), respectively, and incubated at 60°C for 30 min, and the residual enzyme activity was measured according to the standard methods. As seen from Fig. 2.16, the enzyme activity of CGTase was relatively high and stable at pH 6.0–9.0, almost lost at pH 4.0–5.0, and decreased rapidly at pH 10.0 with only 23% of enzyme activity remaining. The pH range of activity for enzyme is relatively narrow compared to the CGTase from other source strain, e.g., the pH of the high activity of CGTase from *B. agaradhaerens* and *B. firmus* is in the range of 5.0–11.4 and 5.5–9.0, respectively.

2.2.3.2.4. Optimum temperature and temperature stability

As shown in Fig. 2.17, the optimum temperature of the CGTase is 60°C in buffer using starch as substrate at pH 6.0 determined by the classic method, and consistent with those of the CGTases from *B. autolyticus* 11149, *B. stearothermophilus* and *B. circulans* E 192.

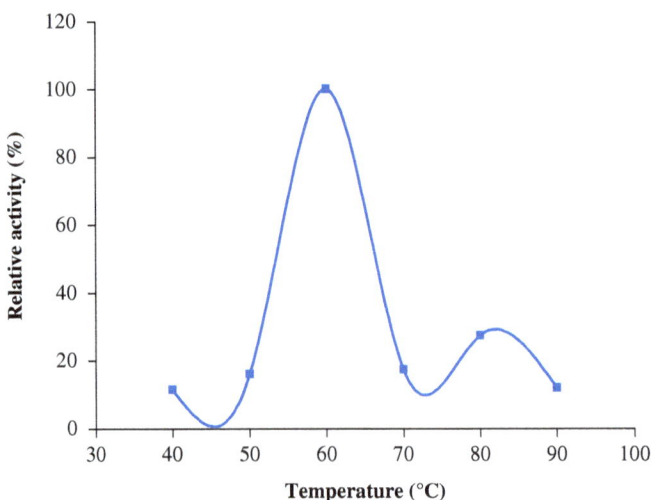

Fig. 2.17.   Effect of temperature on CGTase activity [10].

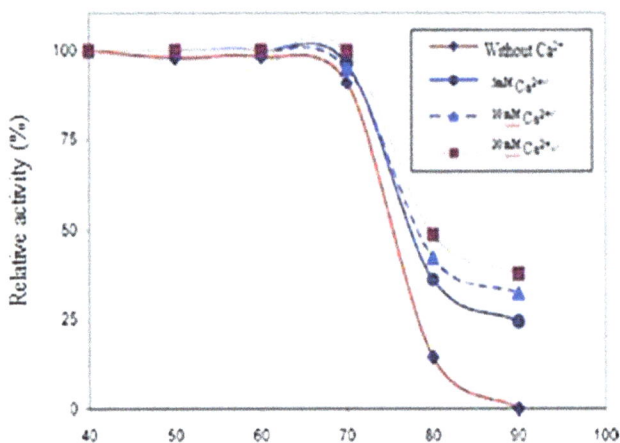

Fig. 2.18.   The thermal stability of CGTase [10].

In the determination of the thermal stability of the enzyme, 0.1 mL of pure enzyme solution was added to 0.2 mL of phosphate buffer (pH 6.0) without any substrate, and incubated at different temperatures (40°C–90°C) for 30 min, and then the residual enzyme activity was determined. As shown in Fig. 2.18, the CGTase activity was maintained relatively high at a temperature of 40°C–60°C, reduced by 50% as the temperature rose to 76°C, maintained only 14% activity as the temperature reached 80°C, and completely inactivated at 90°C. The enzyme from *B. alkalophilus sp.* G1 shows a high thermal stability as compared to the CGTase from *K. pneumoniae* AS-22 and *B. agaradhaeren*. When the calcium ion is present in the reaction system, thermal activity of the CGTase will increase with the increase in the level of calcium ion. When the calcium concentration reached 20 mM, the enzyme could maintain high activity at 70°C, and even retain 37.8% activity when the temperature reached 90°C. This is mainly because calcium plays the role as a protective agent other than just improved reaction rate. Similarily, the thermal stability of the CGTase produced by the *B. circulans* E 192, *B. stearothermophilus* and *Brevibacterium sp.* no. 9605 also increased with the increase in the concentration of calcium.

2.2.3.2.5. Effect of metal ions and other reagents on the activity
          of CGDTase

Effects of metal ions and other reagents on the activity of CGDTase from *B. alkalophilus sp.* G1 was investigated by researchers, and the results are summarized in Table 2.3. Results showed that the activities of CGDTase produced from *Brevibacterium sp.* no. 9605, *B. agaradhaerens, Bacillus sp.* AL-6 and *B. firmus* were significantly inhibited by the copper ions present in the reaction system, while the activities of CGDTase produced from *Brevibacterium sp.* no. 9605, *Bacillus halophilus* INMIA-3849 and *B. firmus, Bacillus sp.* AL-6 were greatly inhibited by zinc ions. These two types of inhibitory effects are due to the oxidation of essential amino acid residues in cyclization reactions.

The enzymic activities of CGDTase and the CGDTase produced from *Brevibacterium sp.* no. 9605 increase in the presence of ferrous ions. In the study of thermal stability of CGDTase, calcium ions have been found to enhance enzyme activity, but there are some exceptions. Activity of CGDTase produced from *B. halophilus*, according to Abelian *et al.*, was not affected by calcium ions [11]. As shown in Table 2.5, activity of CGDTase, with salts composed of metal ion and SO_4^{2-} is higher than that with salts composed of the same metal ion and Cl^-, indicating that anions may impact on the activity of CGDTase.

2.3. Preparation of CGTase by Fermentation

2.3.1. *The types and the expansion of cultivation of the bacteria for preparation of CGTase*

CGTase is produced by microbial fermentation. Good quality bacteria can improve not only the yield of enzyme and utilization of raw materials, but also the enzyme

Table 2.3. Effect of metal ions and other reagents on the activity of CDTase [11].

Metal ions and other reagents (1 mM)	Relative enzyme activity (%)
None	100
$FeCl_2$	189.30
NaCl	72.12
KCl	97.22
$ZnSO_4$	$0.00n$
$CuSO_4$	0.00
$ZnCl_2$	26.79
$MgCl_2$	46.88
$FeSO_4$	0.00
$CoCl_2$	0.00
$NaNO_3$	45.67
$CaCl_2$	106.34
$FeCl_3$	30.00
K_2SO_4	45.54
$MgSO_4$	44.12
$MnSO_4$	57.16
EDTA	17.05
SDS	45.36

Table 2.4. Effect of different carbon sources on CGTase production by *B. firmus* [12].

Carbon sources (10 g/L)	CGTase (U/mL)
Soluble starch	0.33
Cassava starch	0.33
Corn starch	1.05
Potato starch	0.41
Wheat starch	0.59
Cassava starch (Treating with Termamyl^{TM})	0.23
Lactose	0.01

Note: (a) Medium contains yeast extract 5 g/L, peptone 5 g/L.
(b) CGTase activity was not detected when glucose, sucrose, xylose, maltose, sorbitol, mannitol, citrate or glycerol is used as carbon source.

Table 2.5. Effect of different nitrogen sources on CGTase production by *B. firmus* [12].

Nitrogen source	CGTase (U/mL)
Peptone	0.37
Proflo cotton seed powder	0.60
Pharmamedia cotton seed powder	0.69
Casein	0.48
Corn syrup	ND
Soybean powder	0.28

Note: (a) The nitrogen content is equal to 5 g/L the nitrogen content in peptone.
(b) Medium: soluble starch 10 g/L, yeast extract 5 g/L.
(c) ND: CGTase activity not detected.

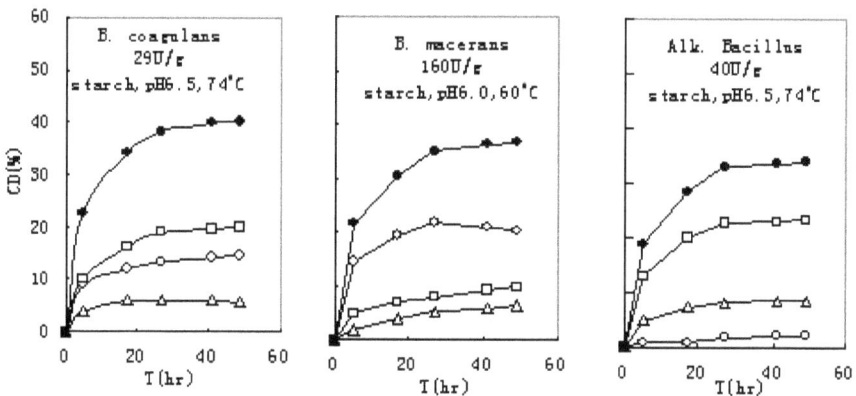

Fig. 2.19. The process curve of starch converted to CD by three kinds of CGTase from *Bacillus* (Potato starch concentration, 13%, w/v; DE < 1.0) [12].

stability and activity. Currently, various types of microorganisms can produce CGTase. The enzymes secreted by these bacteria act on starch or its hydrolysates, and in most cases, all three kinds of α-, β- and γ-CDs can be produced.

However, the proportion of the produced CD by CGTases from varied sources is different. Strains that are commonly used in industrial production are just few species, such as *B. macerans*, *Bacillus coagulans*, *B. alkalophilic*, *B. stearothermophilus*, etc.

Figure 2.19 shows the transformation of starch to CD by CGTase from the three commonly used strains, i.e., *B. coagulans*, *B. macerans* and *B. alkalophilic*.

For CGTase from *B. coagulans,* the yield of CD transformed from starch *is* the highest, in which α-, β- and γ-CDs have a higher content. As to CGTase from *B. macerans,* starch is mainly transformed to α-CD, whereas the yields bond β- and γ-CDs are very low, so it can be used as α-CD production strains. In the case of CGTase from *alkaliphilic Bacillus,* the yields of β-CD and γ-CD are very high, whereas the α-CD production is low, so it is used as β-CD and γ-CD production strains.

In CD production, access to good quality bacteria is the key: first, the bacteria should have a relatively high ability to produce enzyme; second, its enzyme should have the ability to transform starch to the main or unique product of CD. Application of genetic engineering technology to increase the current production of CGTase has been thoroughly carried out. For example, gene cloning method has been used to study enzyme genetic information, gene sequences of DNA, and homology of CGTase produced from different sources of strains, and to reveal the internal determinants of the production specificity of CGTase and the difference between CGTase and α-amylase, etc. The Wacker in Germany introduced cloned CGTase genes into *B. subtilis,* and this strain produced extracellular CGTase 40 times more the original strain. So, with new continuous emergence of research results, access to high yield, thermally stable as well as high specificity CGTase producing strain will become possible, and CD production will also increase in a large scale.

Two types of media are mainly used in microbial fermentation production of the CGTase: one is a common medium for *Bacillus,* i.e., 3% wheat bran (12% corn starch or 1.0% corn dipping sauce) and 0.5% $CaCO_3$ used as a base, to which an appropriate amount of gravy and nitrogen compound such as ammonium sulfate is added; the other is alkaline medium, i.e., 0.5% -1.1% of the carbonate composition added to a base of peptone, yeast extract, K_2HPO_4 and $MgSO_4$. Typically, the fermentation conditions of *Bacillus* strain is: 37°C–40°C, under aerobic incubation, stirring speed of 300 rpm, and fermentation time of 48 h. After fermentation, the fermentation broth is centrifuged, and the supernatant can be used directly as a crude enzyme solution.

In order to ensure a pure fermentation, the seed culture techniques are first used, and then expanded. Industrial use of microorganisms is to use its group effect. According to the law of the general growth and reproduction of the microbial flora, a reasonable training method is used. Strains expansion training is to be aimed at acquiring a large number of strong and simple microbial seeds. Solid slope, eggplant shaped bottle or liquid shake flask are mainly used to culture seeds, the first seed is obtained through the expansion of cultivation. For large-scale production the training can be expanded again. The cultivation of CGTase-producing microorganisms is basically consistent with that of other microorganisms: usually, a bacterial strain is first picked from a cultured slope and added to seed medium, containing corn starch 1%, yeast extract 0.1%, corn steep liquor 5%, K_2HPO_4 0.1%, Mg_2SO_4 $7H_2O$ 0.02% and Na_2CO_3 1% in the liquid medium, dispersed in the cell, and incubated with certain shaking speed and at certain temperatures for 30–48 h, depending on the specific bacteria. After seed culture, the expansion of cultivation and employing the expansion of the medium, similar to the seed medium composition and also known as enzyme production medium, is carried out in the fermentation tank. The inoculum for culture expansion is much larger than that of the seed (generally select 10% of the inoculum), with a necessary set for the speed or ventilation, temperature and incubation time.

2.3.2. *Control of fermentation conditions of CGTase*

Although, in the production of CGTase, the ability of strains to produce enzyme is the determinant factor, the effect of the fermentation process conditions on the production of enzymes is also crucial. Fermentation conditions include not only the composition of the medium, but also temperature, pH, aeration rate, agitation and other factors, and are of interaction and mutual restraint, with appropriate coordination among the said conditions. Therefore, on condition that the production strain is appropriately chosen, the control of the fermentation process is a key factor for the final output.

2.3.2.1. *Medium composition*

2.3.2.1.1. Carbon source

Carbon plays an important impact on the bacterial synthesis rate of CGTase. In the study on the CGTase of *B. cereus* NCIMB13123, it was found that different carbohydrates affect the rate of enzyme synthesis differently, and the synthesis rates from fast to slow were as follows: Xylose > glucose > starch > fructose > galactose > maltose > lactose > sucrose > sorbitol > mannitol > maltodextrin > dextrin. As compared with the starch medium, CGTase enzyme activity increased substantially when xylose and glucose were used as carbon sources in the medium. However, for the production of CGTase by *B. lentus*, if using a single sugar or disaccharide, the activity of enzyme-catalyzed cyclization is 0 [12]. It was also found that the CGTase produced from *B. circulans var. alkalophilus* ATCC 21783 had little cyclization activity grown in the media containing vague essence, soluble starch, β-CD and mannitol medium, and monosaccharide or disaccharide affected enzyme production [12]. *B. stearothermophilus* is able to use lactose, sucrose, glycerol, and sorbitol as a carbon source to produce CGTase. When the concentration of potato starch in the medium reached 10 g/L, *B. circulans var. alkalophilus* ATCC 21783 produced the highest amount of CGTase. Table 2.4 shows the effect of different carbon sources on CGTase production by *B. firmus*.

Typically, metabolites of carbon source in the medium inhibit induction and synthesis of enzyme, and starch is the main carbon source in the medium for CGTase production by fermentation. Take the CGTase production by *B. cereus* NCIMB13123. For example, the production of CGTase with glucose as the sole carbon source is much higher, compared to that with other carbohydrates as carbon sources, and the CGTase synthesis is free from carbon metabolites. Currently, few studies have been conducted on the factors affecting the synthesis of CGTase based on the molecular levels (i.e., the concentration of cAMP, cGMP, ATP, GTP, etc), and transcription and translation (inhibitors of protein synthesis and of RNA polymerase) level.

2.3.2.1.2. Nitrogen source

The CGTase synthesis is significantly affected by the nitrogen sources in the medium, usually in the form of organic nitrogen added to the medium (rarely use inorganic nitrogen source). Corn syrup, yeast extract and peptone are the major sources of nitrogen for CGTase production by bacterial fermentation. Previous studies reported that the hydrolysate of plant protein is the best nitrogen source for CGTase production by *alkalophilic Bacillus spp.* Highest production of enzyme by *B. stearothermophilus* was obtained when defatted soy was used as a nitrogen source [12]. Some bacteria such as basophilic bacteria can use inorganic nitrogen, such as $(NH_4)_2SO_4$, NH_4Cl, NH_4NO_3, $(NH_4)_2HPO_4$, $CHCOONH_3$ and ammonium citrate, etc, but the use of these non-organic nitrogen will not induce the production of CGTase. Among organic nitrogen sources (corn steep liquor, yeast extract, defatted soy, casein, poly peptone, beef extract and sodium glutamate), 5% corn syrup proved to be the best source of nitrogen for production of the CGTase. Carbonates, such as Na_2CO_3 $NaHCO_3$, $CaCO_3$ and K_2CO3, have an important impact on the production of CGTase by alkalophilic bacteria, and the best carbonate (Na_2CO_3 or $NaHCO_3$) concentration in the medium is 1%. Effect of different nitrogen sources on CGTase production by *B. firmus* is shown in Table 2.5.

2.3.2.1.3. Carbonate

Carbonates (Na_2CO_3, $NaHCO_3$, K_2CO_3, $CaCO_3$, etc) in the medium are essential for producing CGTase by basophilic bacteria, and the best carbonate (Na_2CO_3 or

Table 2.6. Effect of carbonate on *basophilic Bacillus* no. 38–2 [12].

Salts	Concentration (%)	Initial pH	Final pH	CGTase (U/mL)
NaOH	—	10.0	8.5	0.79
$CaCO_3$	1.0	7.1	8.6	0.29
$(NH_4)_2CO_3$	1.0	9.2	8.7	0.74
K_2CO_3	1.0	9.9	9.7	0.99
$NaHCO_3$	0.25	8.6	9.3	0.38
$NaHCO_3$	0.5	9.0	9.6	0.84
$NaHCO_3$	1.0	9.5	9.8	1.79
$NaHCO_3$	2.0	9.7	9.9	1.53
Na_2CO_3	0.25	9.4	9.5	0.69
Na_2CO_3	0.5	10.0	9.8	1.76
Na_2CO_3	1.0	10.3	9.8	1.90
Na_2CO_3	2.0	10.7	10.4	0.08
No carbonate	—	7.0	5.9	0.05

$NaHCO_3$) concentration in medium is 1%. Effect of different carbonate on CGTase production by *B. firmus* are shown in Table 2.6.

To obtain a higher yield of CGTase from *Bacillus basophilic* No.38-2, it is better to use Na_2CO_3 or $NaHCO_3$ to adjust the pH to about 9.8.

2.3.2.2. Temperature

In the fermentation process for production of enzymes, there are usually two opposing heat reactions: The first is an endothermic reaction that mainly occurs in synthesis of bacterial cell material and enzymes using the nutrients in the medium by microorganisms in the fermentation process; the second is an exothermic reaction that mainly occurs through catabolism of a large number of nutrients during cell growth. In the initial fermentation, there are less numbers of bacterial cells, and the heat required for the synthesis of biochemical reactions is greater than that for the decomposition of nutrients, and then the temperature of the reaction system at this time shows a downward trend. Therefore, in order to ensure an appropriate temperature required for the growth and reproduction of bacteria and enzyme production, certain insulation measures must be applied. In the latter part of fermentation, due to the large number of cell proliferation, the heat released during decomposition reaction is greater than that absorbed, so fermentation temperature will rise on its own, combined with the mechanical heat released in ventilation and mixing, so the fermentation system must be cooled. The temperature is usually maintained at 37°C–40°C in the production of

CGTase by *Bacillus* genus. Fermentation temperature. 30°C was reported in some literatures, depending on the balance between the growth of bacterial activity and enzyme production activity. In actual production, reasonable adjustments should be made based on the selected bacteria.

2.3.2.3. *pH*

Microbial growth and reproduction require a certain pH environment. CGTase-producing *Bacillus* bacteria are of no exception, and if the pH value is not suitable, the normal cell growth and reproduction are hindered, and the microbial metabolism pathways and the nature of metabolites also change. In the CGTase production, the pH of the seed medium and fermentation directly affect the production and quality of the enzyme. For the process of microbial fermentation for enzyme production is the actual process of biochemical reactions of growth metabolism of microorganisms, pH of the reaction system constantly changes. In general, if the C/N ratio of the medium composition is high, the fermentation broth exhibits acidity and the pH is low; while if the C/N ratio of the medium composition is low, the fermentation liquid shows alkaline and the pH is high. Ventilation affects the pH to a certain extent mainly because it affects the level of oxidation of the sugar and fat in the medium, and when there is incomplete oxidation, the system will generate intermediate organic acids, making the system pH a downward trend. In the fermentation process, the use of nitrogen by microorganisms can also lead to changes in the pH. In short, the variety of reactions will influence changes in the pH, and such reaction pH in different periods are often used to regulate actual production and fermentation. In the production of CGTase, the pH control is usually done by adding buffer to adjust the pH of the reaction system and to maintain a certain pH range. In practice, a phosphate buffer can be chosen mainly from K_2HPO_4, KH_2HPO_4 and Na_2HPO_4, or mixtures of them. As the source of CGTase-producing microorganisms is different, the pH requirement for each species is also different, i.e., a number of basophilic bacteria require high pH, while some other bacteria require nearly neutral pH. Therefore, on the choice of bacteria, the first thing to consider is the fermentation of the pH, and according to the actual requirements, the initial pH of the medium and the pH of the fermentation process are to be adjusted.

2.3.2.4. *Ventilation*

CGTase-producing microorganisms are basically aerobic, and during the processes of growth and reproduction, most still require oxygen uptake from the reaction system. Therefore, providing an appropriate oxygen component helps improve the enzyme production from such microorganisms. Of course, too much oxygen is not better than appropriate oxygen, and a variety of microorganisms require the existence of a critical oxygen concentration, and too much or too little oxygen will inhibit microbial growth and normal metabolism, thereby changing the metabolic pathways and affecting the types or numbers of metabolites generated. In initial fermentation, due to a small number of bacteria, ventilation can be maintained at a low level; in active enzyme producing fermentation period, cell growth and reproduction are also active and demand more ventilation, but to some bacteria, vigorous ventilation could retard enzyme production. In the laboratory, flask fermentation is usually the way to ventilation. In the laboratory A. Rosso *et al.* used the conditions 37°C and 100 rpm shaker fermentation to prepare the CGTase [13]. Cao Xin-Zhi found that when the liquid volume increased or the speed decreased, the enzyme activity produced by the bacteria was reduced. At lesser liquid volume or higher speed, the enzyme activity for the CGTase production by *alkalophilic Bacillus* was higher. This is mainly because *alkalophilic Bacillus* are aerobic, and when the liquid volume decreased or the speed increased, adequate amount of oxygen was supplied to the system. The increase in dissolved oxygen is more conducive to bacterial growth, which also increased the enzyme activity [12].

In summary, different culture conditions are required for different microorganisms for the production of CGTase (i.e., different medium composition, pH and temperature for the optimum condition). The culture conditions for CGTase production by some important bacteria are summarized in Table 2.7.

2.3.2.5. *Immobilized production of CGTase*

There are generally three fermentation ways for the CGTase production system: Free or immobilized cells in batches, repeated batches, and continuous culture. The CGTase production systems can be divided as solid-state fermentation and submerged fermentation. Usually submerged fermentation is used, for it can precisely control the ventilation, stirring rate, fermentation temperature and pH and other parameters, and also maintain a high degree of sterility. Kunamneni *et al.* used sodium alginate immobilized cells of *alkalophilic Bacillus sp.* fermentation and non-immobilized fermentation to produce CGTase in air-lift reactor, and found that the yield of the immobilized continuous production was $13.65\,UmL^{-1}h^{-1}$, while that of the direct fermentation was only $6.89\,UmL^{-1}h^{-1}$ [14]. In the air-lift reactor, immobilized *Bacillus sp.* has a longer half-life, and can be subjected to continuous fermentation upto 18 days. In the study of CGTase production by immobilized cells carried out in batches, semi-continuous, and continuous culture, the cells of *B. circulans var. Alkalophilus* ATCC21783 were embedded in

Table 2.7. The culture conditions of CGTase by part of important bacteria [12].

Bacteria	Optimum pH	Optimum temperature (°C)	Composition of medium	Time of maximum enzyme production (h)
B. macerans ATCC 8514	7.0	40	Starch, casein hydrolysates, yeast extract	10~15
B. macerans IFO 3490	7.0	37	Soluble starch, corn syrup, $(NH_4)_2SO_4, CaCO_3$	70
B. megaterium No.5	7.0	37	Water-soluble starch, wheat bran, poly peptone, corn steep liquor, dry yeast	70
B. circulans var. alkalophilus ATCC21783	10.3	37	Soluble starch, corn syrup, K_2HPO_4, $MgSO_4, Na_2CO_3$	48~50
Bacillus sp. AL-6	10.0	38	Soluble starch, corn syrup, K_2HPO_4, $MgSO_4, Na_2CO_3$	48
B. cereus NCIMB 13123	7.0	37	Glucose, peptone, $K_2HPO_4, MgSO_4$	16~20
B. halophilus INMIA-3849	7.2	37	Potato starch, peptone, yeast extract, sodium citrate, NaCl, KCl, $MgSO_4$	48
Bacillus ohbensis sp. Nov. C-1400	10.0	30	Water soluble defatted soy, sodium glutamate, K_2HPO_4, $MgSO_4, Na_2CO_3$, $NaHPO_4, CaCl_2$	72

calcium alginate gel. The cell's density was adjusted within the gel ball and with medium composition during the batch fermentation, reusable immobilized cells produced enzymes with activity of about 60 U/mL in each batch fermentation. They were continuously cultured in fluidized bed and CGTase activity was up by 40–50 U/mL. Such activity can be maintained for 450 h. Another study was sodium alginate immobilized cells of *Bacillus cereus NCIMB* 13123 for batch fermentation, where the maximum enzyme activity was 40 U/mL after 24 h, and then due to the consumption of nutrients inside the gel ball, the enzyme activity drastically reduced. Therefore, the culture medium needed to update every 24 h, and after 10 repeated batch fermentations, CGTase activity still remained at 40 U/mL level. For continuous production of CGTase, the immobilized cells were cultured in fluidized fixed-bed for 400 h, where the reactor dilution rate varied within the first 120 h and was maintained at $0.32\,h^{-1}$, and then the enzyme activity was stabilized at 60 U/mL after 280 h. Practice has proved that the highest enzyme activity of CGTase produced using immobilized cells for continuous production was 10% to 40% greater than the free cell batch or continuous production, and the enzyme production using continuous system was 5 to 13 times higher than that using other fermentation systems. Therefore, it is preferable to use immobilized cell continuous fermentation for large-scale production of CGTase.

2.3.3. *Determination of CGTase activity*

The so-called enzyme activity refers to the enzymatic catalysis ability of a chemical reaction. Therefore, enzyme activity can be expressed as the catalytic rate of a chemical reaction under certain conditions: high reaction rate means great enzyme activity, and vice versa. Similar to general chemical reaction, the enzymatic reaction rate can be represented by the reduction in the substrate or increase in the amount of product per unit time and per unit volume. In general, determination of enzyme activity is based on the increase in the product. According to the determination principle, enzyme activity can be divided into four categories: (a) End-point method, i.e., the reaction is stopped at a certain time, and then chemical or physical changes in the product or the amount of substrate are determined; (b) Kinetic method, i.e., without sampling and the termination of reaction, the enzyme reaction product, substrate or coenzyme of variation are continuously measured during enzymatic reaction, and the initial velocity of enzyme reaction can be directly measured; (c) Enzyme-coupled assay, i.e., excessive highly specialized "coupling tool enzyme" is added to the measured enzyme reaction system, and the reaction proceeds to a direct, continuous and accurately measured stage; (d) Electrochemical method, i.e., ion-selective electrode is used to track the concentration of the ion or gas molecules generated in reaction, resulting in the initial reaction rate.

The determination of CGTase activity has been in a progressive state of exploration because the enzyme has four catalytic activities and it is difficult to put them together in the determination of normal activity. In CD production catalyzed by CGTase, four types of reactions, i.e., the cyclization, hydrolysis, coupling and disproportionation, take place simultaneously. Because there are several components in the system at the same time, the CGTase activity calculated simply by the determination of α-amylase, i.e., determination of substrate consumption using iodine color method is not accurate. Nakamura and Horikoshi improved this method: 0.05% amylose phosphate buffer and 0.2 mL enzyme solution were incubated at 40°C for 10 min. To the reaction solution 4 mL 0.2 mol/L HCl, 0.5 mL iodine, 4 mL of distilled water was added, and the absorbance was measured at 700 nm. An activity unit is defined as the amount of enzyme that decreases starch–iodine complex per minute by 10% [15]. A similar approach was developed where 3% amylose (w/v) was prepared using 5×10^{-4} mol/L $CaCl_2$-phosphate buffer, and the pH was adjusted to 6.2 with 1 mol/L HCl. To 5 mL of this substrate solution, 1 mL CGTase solution was added and incubated at 40°C for 30 min. Then, 0.2 mL of this reaction mixture was introduced into the 100 mL flask containing 0.2 mL iodine solution, and diluted to scale using 1 mL 1 mol/L HCl, and then the absorbance was measured at 620 nm (AbD). When the same volume of distilled water (instead of enzyme solution) was mixed with HCl, substrate solution and iodine mixture, diluted to 100 mL, and used as the blank, the absorbance was measured at 620 nm (AbC).

$$\text{Apparent enzyme activity (units/mL)} = [\text{AbC-AbD/AbC}] \times 125$$

The activity calculated according to above formula includes both the enzyme activity of dextrin and the generation of activity.

A semi-quantitative analysis approach to the activity of synthetic CGTase has now become a CD synthesis activity (CS activity). There is a determination the preparation of continuous double dilution of CGTase. 0.5 mL enzyme solution and 0.5 mL of substrate are incubated at 40°C for 30 min. One drop of HCl is added to stop the reaction. A drop of the digestive enzyme is blended with one drop of iodine solution on the slide, and the crystallization at droplet edge is observed. CS activity can be expressed as the reciprocal of the highest dilution times of CGTase that can be detected. With microscopy, the blue–black α-CD-I2/KI crystal generated is easy to observe, while β-CD-I2/KI yellow–brown needle-like crystals are sometimes difficult to distinguish. The method using inclusion complexes of β-CD and PP to decrease absorbance at 555 nm can quantitatively determine the CGTase activity of the ring. For the quantitative determination, starch solution (40 mg/mL) was prepared using 0.1 mol/L sodium phosphate buffer (pH value of 6.0), mixed with 0.1 mL β-CGTase, appropriately diluted with distilled water, incubated at 60°C for 20 min, and finally, NaOH and PP were added to stop the reaction. By plotting the decreased percentage (%) of absorbance against the [CD], the linear relationship is obtained and used as the working curve (a straight line when the conversion of α-CD is less than 2 mg). An enzyme unit is defined as the amount of enzyme required to generate 1 mg β-CD each minute under the experimental conditions.

The HPLC method to determine the amount of CD generated is fast and accurate in determining the activity of CGTase. About 0.5 mL CGTase solution and 6% starch solution (MacIlvaine Buffer 100 mmol/L, pH 6) 2.5 mL were incubated at 50°C for 60 min, boiled to inactivate the enzyme activity, and separated on NH_2 or ODS column. There is a linear relationship between the resulting amounts of α-, β- and γ-CDs (sample volume of 5 μg) and amount of β-CD (mg) when CGTase concentration is below 10×10^{-2} units.

2.4. Purification of CGTase

2.4.1. *Pretreatment and filtration of fermentation broth*

In the process of CGTase production by fermentation, after the termination of fermentation, in addition to enzymes and other metabolites, the fermentation broth also contains large amounts of bacteria and medium residues. The purification of the enzyme cannot be completed without filtration and separation processes of the broth. If the CGTase prepared without this process is directly applied to the production of CD, the yield and quality of CD would greatly be reduced. Thus, the broth must be processed to obtain clarified enzyme solution. In the preparation of CGTase, bacteria and some medium residues are usually removed by centrifugal separation.

2.4.1.1. *Salting*

The neutral salts commonly used in salting are $MgSO_4$, $(NH_4)_2SO_4$, Na_2SO_4 and NaH_2PO_4. In the concentration of CGTase, $(NH_4)_2SO_4$ is generally used as a salting agent, mainly because at lower temperature, its solubility is greater (its solubility is 706 g/L at 0°C, and 767 g/L at 25°C), and CGTase preparation process is mostly carried out at low temperature. Therefore, enzyme preparation by $(NH_4)_2SO_4$ salting has the advantage of not resulting in inactivation of the enzyme, and non-enzymatic protein precipitation entrained in the precipitation process is less. However, due to the bad smell of $(NH_4)_2SO_4$, this crude product cannot be directly applied in the food industry, and other follow-up processes must be combined to obtain quality standards of CGTase. Cao Xin-Zhi *et al.*, purified CGTase produced by *alkalophilic Bacillus sp.*, using ammonium sulfate, DEAE-cellulose chromatography and Sepharose CL-6B chromatography column to treat the crude enzyme solution [12].

2.4.1.2. *Organic solvent precipitation*

In order to concentrate the crude enzyme solution, organic solvent precipitation method can be used to to produce CGTase from *basophilic Bacillus Bacillus* no.38-2 (ATCC 21783): three volumes of acetone (pre-cooled at $-20°C$) were added to the crude enzyme solution at $0°C$, then the mixture was placed at low temperature overnight, and collected by filtering sediment.

2.4.1.3. *Adsorption*

Due to the diversity of strain sources, the choice of filtering methods is also different. In the preparation of CGTase, fermentation broth is usually pretreated by centrifugation. For example, the fermentation broth was centrifuged at 3,000 r/min for 15 min, and crude enzyme from the supernatant was obtained by Cao Xin-Zhi [12]. Similarly, in some other preparation processes, the cells were removed by filtration, i.e., in the production process of CGTase produced from *B. megaterium*, the cells were removed by filtration with diatomite; in the case of CGTase production from *Alkaliphilic Bacillus* no.38-2 (ATCC 21783), $CaCl_2$ was added to the broth with stirring at $0–4°C$ to maintain the pH of 7–8, and supernatant was obtained by centrifugation. In the preparation process of CGTase, operation is usually carried out at $0–4°C$ mainly because of the great relationship between the effect of enzyme precipitation and temperature: in general, the lower the temperature, the more precipitation is obtained. Part of the precipitated enzyme mixture will be dissolved again if the temperature rises. New deposits are likely to happen in the supernatant obtained by filtration due to the decrease in temperature. Therefore, in the processes of sedimentation, filtration or centrifugation, the temperature should be strictly controlled in a certain range to avoid wide temperature fluctuations.

2.4.2. *Concentration and precipitation of crude enzyme solution*

The impurities and the low amount of enzyme in the crude extract leads to a low catalytic efficiency of the enzyme solution. For further purification, the crude extract must be concentrated, and the enrichment processes commonly used are evaporation, ultrafiltration, gel filtration, repeated freezing and thawing and freeze-drying.

2.4.3. *Enzyme purification*

2.4.3.1. *Affinity chromatography*

This approach had been used an the preparation of CGTase as early as the 1930s, and corn starch was used as adsorption by the formers, but there was no affinity chromatography method at that time. With development of scientific research, and using α-CD or β-CD as adsorbents, CGTase can be adsorbed more specifically. In the separation of CGTase from broth of *Paenibacillus sp.* F8, Rha *et al.* used β-CD as ligands, agarose gel 6B as the carrier (hydroxyl group coupled) and α-CD solution as the eluent for affinity chromatography [14]. α-CD solution was selected instead of β-CD or γ-CD solution as the eluent, because α-CD does not interfere with the accuracy of protein measured using Bradford method.

2.4.3.2. *Ion-exchange chromatography*

The isoelectric points (pI) of most CGTases are about 9, and the pK of DEAE is 9.5, so the DEAE-cellulose column, using DEAE as the exchange agent, is mostly used for the separation of CGTase. This method is often used in the previous preparation process.

2.4.3.3. *Gel filtration or gel permeation chromatography*

According to the molecular weight range of CGTase, in the gel separation of the enzyme, Sepharose and models 6B are mostly used. In many research reports, combination of adsorbent with gel chromatography was mostly used: adsorbents such as starch, α- and β-CDs etc were coupled in Sepharose column to form affinity chromatography. In the purification of CGTase, Rha *et al.* used β-CD and Sepharose 6B [16].

2.4.3.4. *Dialysis*

Dialysis membranes can be animal membranes and cellophane, but mostly used membranes are made of cellulose. Now, various-sized dialysis tubes, made by the American Union Carbide and American medical spectrum, are commonly used. The MWCO of the tubes are usually around 10,000. In order to improve the efficiency of dialysis, a variety of devices can also be used, including various types of Zeineh dialyzer (Biomed Instruments Inc. US), by which the speed and efficiency of dialysis can be greatly increased.

In the preparation of CGTase, dialysis can often be used for desalination. Most of the preparation processes use ammonium sulfate precipitation method, and some of the preparation processes use buffer solution such as Tris-HCl buffer or acetate buffer fluid, introducing a number of ions to the system. In order to obtain pure CGTase, dialysis approach is mostly employed to remove salts, which is an effective way.

2.5. The Substrate Catalysis Characteristics and Sources of Pullulanase

2.5.1. *The substrate catalysis characteristics pullulanase*

Pullulanase (EC3.2.1.41) is a debranching enzyme that specifically cut-off α-1,6-glycosidic bond in pullulan, starch, amylopectin and the corresponding oligosaccharides. In the starch industry, commonly used α-amylase and β-amylase only act on the α-1, 4-glycosidic bond and cannot work on the α-1, 6-glycosidic bond in starch. Pullulanase has the ability to cut-off the α-1, 6-glycosidic bond and has attracted wide attention since its hydrolysis activity was found. Some of pullulanases are the acidic enzymes, where the optimum pH is usually 4.0 to 6.0, and the activity is unstable under alkaline condition. Alkaline pullulanase not only has a relatively high stable enzyme activity at pH 6.0–11.0, but also has high thermal stability. Under alkaline conditions, it can specifically hydrolyze the α-1,6-glycosidic bonds in pullulan, starch, glycogen, amylopectin and the corresponding oligosaccharides, and can significantly improve the utilization of starch as raw materials. The combination of pullulanase, α-amylase and β-amylase has been applied to produce high glucose syrup, high maltose syrup and oligosaccharide.

According to substrate specificity and different decomposition of products, pullulanase can be broadly divided into four categories:

(1) Pullulanase Type II (glucoamylase EC3.2.1.3): Hydrolysis of α-1, 4-glycosidic bond from the non-reducing end of pullulan to generate glucose.
(2) Pullulanase Type I (Pullulanase EC3.2.1.41): Hydrolysis of α-1, 6-glycosidic bond in pullulan and straight-chain oligosaccharides to generate maltotriose and linear oligosaccharides.
(3) Pullulanase Type I (neopullulanase): Hydrolysis of α-1, 4-glycosidic bond in pullulan to generate Pan sugar.
(4) Pullulanase Type II (isopullulanase EC3.2.1.57): Hydrolysis of α-1, 4-glycosidic bond in pullulan to generate α-maltose-based-1, 6-D-glucose.

Mg^{2+}, Ca^{2+}, Mn^{2+} are the activators of pullulanase; Hg^{2+}, Cu^{2+}, Al^{3+}, Zn^{2+}, etc inhibit enzyme activity; EDTA and CD etc also have a certain extent of inhibition on the enzyme; Ca^{2+} can effectively improve the pH stability of enzyme and thermal stability.

Pullulanase, together with β-amylase, can act on amylopectin or glycogen to generate almost 100% maltose. It can hydrolyze amylopectin, but cannot solely hydrolyze more densely branched glycogen may be due to the steric hindrance of intensive glycogen branches, which makes the enzyme difficult to reach sites of action. When combined with β-amylase, due to trimming-off part of branched-chain by β-amylase, α-1, 6-glycosidic bonds are exposed. Under the conditions of the high concentration of substrate and high enzyme concentration, pullulanase can also catalyze the condensation reaction of maltose and maltotriose to form the branching maltotetraose, maltopentaose and maltohexaose. The maltotetraose produced through the condensation reaction by the pullulanase from *Bacillus* can be hydrolyzed by glucoamylase to generate glucose and panose.

When the pullulanase from bacteria and plant acts on α-1, 6-glycosidic bonds in pullulan, the reaction first generates maltotriose and a series of branches of the maltotriosyl-oligosaccharides by dissociation, and the final product is a maltotetraose. In the study of hydrolysis of the various branches of oligosaccharides and limiting dextrins, it was found that it can hydrolyze the α-1, 6-glycosidic bond at the point of collateral branches composed of 2 to 3 glucose residues in α- and β-limiting dextrin, but for the oligosaccharides such as panose, isomaltose, isopanose or polysaccharide that only has one glucose residue linked to the α-1, 6-glycosidic bond, pullulanase cannot work, in other words, the α-1, 6-glycosidic bond to be cut-off by pullulanase has to have at least two or more α-1, 4-glucose units (Fig. 2.20).

2.5.2. *Source of pullulanase*

Pullulanase was obtained originally from *Aerobacter aerogenes* fermentation by Bender and Wallenfels in 1961. In 1975, Yoshiyuki Takasaki found that the

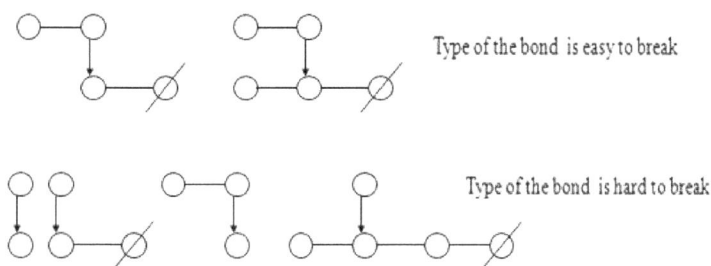

Fig. 2.20. The specificity of pullulanase of α-1, 6-glycosidic bond [18].

B. cereus var. Mycoides could produce two kinds of starch enzymes: β-amylase and pullulanase, where the optimum condition pH is 6–6.5, temperature is 50°C, and the maximum conversion rate (maltose from starch by hydrolysis) is about 95% [18]. In the 1980s, Novo Nordisk Denmark had received *Acidophilic Bacillus* that hydrolyzed pullulan; the pullulanase from it is now the most widely used, and has the largest output [18]. In 1986, Yoshiyuki Takasaki isolated *B. subtilis* producing heat and acid stable pullulanase, which could produce the mixture of pullulanase and amylase, of which the optimum pH of pullulanase was 7.0–7.5, but also maintained 50% of enzyme activity at pH 5.0 [19]. In 1987, E. Madi and G. Antranikian reported a simultaneous production of α-amylase, pullulanase and glucoamylase bacteria: *Clostridium thermosulfurgenes*. In addition, some *actinomycetes* such as *Streptomyces diastatochromogenes*, *Beauveria actinomycetes* and *Micromonosporaceae, Actinomycetes thermomonosporaceae* also produce pullulanase. In plants, such as rice, beans, potatoes, sweet corn and malt, pullulanase was observed [20].

2.6. The Substrate Catalysis Characteristics and Sources of Isoamylase

2.6.1. *The substrate catalysis characteristics and of isoamylase*

Isoamylase (EC3.2.1.68) and pullulanase belong to the debranching enzymes, and the difference is that isoamylase cannot hydrolyze pullulan, but has high ability to hydrolyze densely branched amylopectin and glycogen. The best substrate of isoamylase is a polysaccharide polymer, while that of pullulanase is the polysaccharide with low molecular weight. This difference reflects that the isoamylase needs strict specificity of true branching structure. Similar to the amylase from plant, microbial isoamylase can specifically hydrolyze the α-1, 6-glycosidic bonds in starch and glycogen, resulting in amylose and dextrin. In the starch industry, isoamylase combined with external enzymes, acts on starch to generate glucose, maltose and some oligomeric sugars; at the same time, isoamylase can transform branches into CD, and improve branching CD solubility by the enzyme reverse function.

Isoamylases of different microbial sources have a greater difference in protein and enzyme characteristics. For example, yeast isoamylase can significantly increase the hydrolysis efficiency on amylopectin and glycogen with the α-amylase, and together with β-amylase, it can completely hydrolyze amylopectin and glycogen to generate maltose, but it has no effect on pullulan and glucan. As a preliminarily produce, the enzyme was purified by isopropanol precipitation and carboxymethyl cellulose chromatography. The optimum pH and temperature were 5.6 and 30°C, respectively, with a molecular weight of about 65,000, and isoelectric point of 4.7–4.8. Isoamylases are inhibited by Hg^{2+}, Cu^{2+}, Fe^{3+}, Al^{3+} and other metal ions, and maltotriose, maltotetraose, amylose, corn amylose and other carbohydrates also strongly inhibit the isoamylase activity. Mg^{2+} and Ca^{2+} have some activation effects on the enzymes, and dextrin is a good inducer for the enzymes. Therefore, the production of straight chain dextrin and high maltose syrup from amylopectin and glycogen using isoamylase has certain advantages as compared with that using pullulanase; but due to the instability, its application in industrial production has a great limitation.

2.6.2. The source of isoamylase

Many microorganisms are capable of producing isoamylase: yeast, *B. licheniformis*, *B. cereus*, *Bacillus amyloliquifaciens*, *Flavobacterium*, *Pseudomonas*, etc, can secrete isoamylase. The isoamylase from the *Pseudomonas* isolated from soil was purified, and the nature and function of the enzyme were studied in detail. Pullulanases from *Pseudomonas* and *Bacillus aerogenes* were characterized for enzyme kinetics and the effects on glycogen and amylopectin were compared by Kubo *et al.* [21]. Sugimoto *et al.*, through traditional means of mutation breeding obtained a mutant with higher activity, and successfully applied the enzymes to industrial production. With the continuous development of genetic engineering techniques, the use of cloning technology to improve the activity of isoamylase also has made substantial progress.

2.7. The Source and Nature of Galactosidase

Galactosidase, also known as galactosidase hydrolase, includes both the α-galactosidase, and β-galactosidase, and some of them can be used for the transfer of galactosyl to prepare branched-CDs. However, the source, substrates and enzymatic properties of these two enzymes vary greatly and are introduced as follows.

2.7.1. α-Galactosidase

2.7.1.1. The source of α-galactosidase

α-galactosidase (EC3.2.1.22), also known as melibiase, can specifically hydrolyze the α-galactosidase bond (Fig. 2.21). This enzyme can hydrolyze not only some simple α-galactosyl bonds, but also complex α-D-galactosidase molecules, such as oligosaccharides, polysaccharides, raffinose, raffinose, stachyose and verbascose as well as some heteropolysaccharide containing galactosyl bonds.

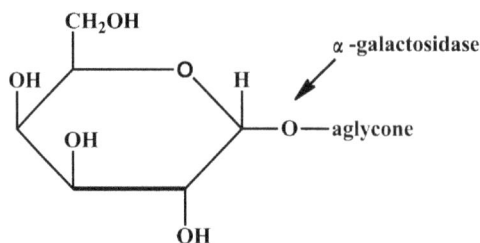

Fig. 2.21. The hydrolysis function of α-galactosidase [23].

Table 2.8. α-galactosidase source [23].

Plant	Animal	Microoganism
Guar and its gum, coffee beans, lentils, beans, green beans, melon, watermelon, clover and sugar cane	Rat liver, human plasma, human placenta, snails, etc	*B. cereus, E. coli, B. stearothermophilus, Lactobacillus, Candida, Pichia pastoris, Aspergillus, Penicillium, Aspergillus, Mucor, Streptomyces Axl Rose*

α-galactosidase was found in the early 20th century, but it has always existed in nature. The source of this enzyme is very wide, and exists in seeds and tissues of mammals, and many microorganisms can also produce α-galactosidase. Table 2.8 lists the various α-galactosidase source.

Different sources of α-galactosidases show different catalytic activities, and the sites of hydrolyase and the glycosyltransferase are relatively different. Take the catalysis of the synthesis of the glycosidic bond. For example, different sources of glycosidases have great differences in the regional selectivity, and form glycosidic bonds on the hydroxyls at different locations of the same sugars in glycosides. The α-galactosidases used in the preparation of branching CDs are mainly produced from green coffee beans, while the α-galactosidases from *Mortierella vinace* and other microoganisms can also be used in the preparation of some branching CD. Different sources of α-galactosidases have different molecular weights (23,000–325,000), and most of these different sources of α-galactosidases have a variety of molecular forms: some are apparently polymer forms composed by the same kinds of subunits, such as that from soybeans; others have similar but significantly different molecular weights, such as those from *Medicago sativa*, which may be due to the differences in the basic structure of protein, or in the glycosylations combined. In different stages of seed germination, different forms of the enzyme can be transformed into each other. The optimum pHs of most α-galactosidase enzymes vary in the range of 4 to 7, and if the substrates are different, the optimum pHs will change accordingly: When the substrate is the artificial pNPG, the optimum pH of α-galactosidase from coffee beans is 6.4, and when the substrate is melibiose, raffinose or stachyose, the optimum pH drops to 3.6–4.

2.7.1.2. *The transferase activity of α-galactosidase*

For the α-D-galactosyl bonds, α-galactosidases from different sources show wide substrate specificity. The α-galactosidase from the coffee bean hydrolyzes not only low molecular weight substrates (such as pNPG, melibiose, etc), but also the high molecular weight oligosaccharides (such as starchyose) and polysaccharides (such as galactomannan), and can also act on the B-antigens of human body. However, different sources of α-galactosidases show differences in the substrate specificity. Most of the α-galactosidases from eukaryotic cell show similar specificity for the low molecular weight color source substrates (such as pNPG), and also have differences in their activities for high molecular weight oligosaccharides and sugar combination, which may be due to stereochemistry of enzymes and substrates. So far, only several α-galactosidase reported can hydrolyze the α-galactosyl bonds of type B red blood cells. They were produced from beans, soybeans, coffee beans, etc. α-galactosidases show very high specificity for the stereochemistry of glycosidic bond and sugar moiety, and have no activity for β-D-galactosidase and α-DX glycosidic bond (X is non-galactose).

For most of the α-galactosidases, low concentrations of metal ions such as Ag^+, Hg^{2+}, Cu^{2+}, can somewhat inhibit the enzyme activity, and the inhibition mechanism of these metal ions is relatively complex: they can act on not only thiol, but carboxyl or imidazole and so on. Substrates of α-galactosidase, such as melibiose, tend to be inhibitor of α-galactosidase. Here is the order of strength of the inhibition by several galactosidases of *Vica faba*: galactose > raffinose > melibiose > stachyose. When the *V. faba* seed germinates, D-galactose produced by hydrolysis of oligosaccharide strongly inhibits the α-galactosidase activity by inhibiting utilization of the source from the oligosacchrides reserves. But the seed can transform galactose into galactose-1-, which is the weak inhibitor of α-galactosidase. D-galactose and its structural analogues, such as arabinose and fucose, can also inhibit α-galactosidase hydrolysis activity. In the presence of D-galactose, the inhibition effect of mercury chloride acid (PCMB) on α-galactosidase will be with a certain degree of ease, and this may be due to a competitive combination of active sites of D-galactose and α-galactosidase.

α-galactosidase can catalyze the hydrolysis of α-galactosyl bond of the pNPG melibiose and other donors of galactosyl, and transfer galactosyl in the presence of

Table 2.9. α-galactosidase enzyme receptor specificity of transfer [23].

Sources	Glycosylated receptor	Transformation product
Monosacchrides	Glucose	Melibiose
	Galactose	Galactobiose, Galactotriose,
	Fructose	Galactotetraose
		None
Disacchrides	Sucrose	Raffinose
	Melibiose	Manninotriose
Oligosacchrides	Raffinose	Starchyose, Verbascum Sugar,
		Sugar Ajuga

high concentrations of glycosylated receptor. Different sources of α-galactosidases have differences in the rate of hydrolysis of glycosidic bonds and transfer of galactosyl and receptor specificity. The α-galactosidases from *M. vinace* and other microorganisms exhibit transfer activity on glucose and maltosyl-CD, while there is no activity on the CD and glucosyl-CD. The α-galactosidase from coffee beans has higher transferase activity and more extensive of receptor specificity.

Malhotra and Dey studied the receptor specificity of the α-galactosidase transferase in the sweet almond, and the results are shown in Table 2.9. They found that the strength of the inhibition of hydrolysis of α-galactosidase by three monosacchrides and their receptor specificity is the same: galactose > glucose > fructose, which indicates that transfer reactions may be related to the degree of integration of receptor sites and enzyme activity, and when galactose and glucose with the same concentration co-exist, only glycosyltransferase products are produced from galactose because the glucose is a weak glycosylated receptor as compared with galactose. Malhotra and Dey also believe that α-galactosidase catalyzes the hydrolysis and glycosyl transfer reactions by the same active site, and has the same mechanism, and in the reaction process, the hydrolysis of the donor and the formation of products occur simultaneously [23].

2.7.2. β-galactosidase

2.7.2.1. Source of β-galactosidase

β-galactosidase (EC3.2.1.23), also known as lactase or β-D-galactoside galac-tohydrolase, is an important glycoside hydrolase. The enzyme can hydrolyze galactosidase bonds to generate a mixture of galactose and glucose, and some

Table 2.10. The source of β-galactosidase [24].

Source	Name
Plant	Almond, coffee beans, alfalfa, corn, cycads, soybean, sesame, apricot, apple
Animal	Brain tissue, intestinal tissue, skin, intestinal tissue especially of the baby
Fungal	*Aspergillus niger, Aspergillus flavus, A. oryzae, Okinawa Aspergillus, Mucor, Alternaria mold, sky blue Streptomyces*
Yeast	*Utilis yeast, lactic acid Kluyveromyces, Kluyveromyces fragilis*
Bacterial	*B. megaterium, B. stearothermophilus, E. coli, Lactobacillus bulgaricus, Streptococcus lactic acid, Streptococcus thermophilus, etc*

Table 2.11. The nature of the β-galactosidase [24].

Name	Optimum pH	Stability arrange of pH	Optimum temperature (°C)	Molecular weight (KD)	Impact factor	Lactose Km
A. niger	3.0–4.0	2.5–8.0	55–60	124	—	85
A. oryzae	5.0	3.8–8.5	50–55	90	—	50
K. fragilis	6.5	6.5–7.5	37	201	$Mn^{2+}K^+$	14
K. Lactis	6.9–7.2	7.0–7.5	35	135	$Mn^{2+}K^+$	12–17
E. coli	7.2	5.0–8.0	40	540	Na^+K^+	2
Lactobacillus thermophilus	6.2	—	55	540	—	6
B. stearothermophilus	6.5	—	60–65	—	—	2.7

β-galactosidases can transfer glycosides: transfer galactosidase to lactose and produce galactooligosaccharides.

The natural sources of β-galactosidase are very wide: leaves, stems and seeds of plants (such as almond, corn soybeans, etc); animal skin and intestinal tissue, etc; microorganisms, including fungi, yeasts and many species of bacteria. The primary sources of β-galactosidase are listed in Table 2.10.

Among the sources listed in Table 2.10, β-galactosidases from microbials are the most commonly used in industrial purposes because of wide source, higher yield and low production costs. Currently, the main products are produced from *A. niger, K. lactis, K. fragilis* and *E. coli*.

Different sources of β-galactosidase have different enzymatic properties. The different nature of β-galactosidase with different optimum pHs and temperatures of enzymes determine their different uses, as shown in Table 2.11.

2.7.2.2. *The transferase activity of β-galactosidase*

The transfer function of β-galactosidase and the hydrolysis process occur simultaneously. β-galactosidase hydrolysis can usually be done in two steps:

The first step: β-galactose-1,4-α-glucose + β-galactosidase → complex (β-galactosidase - galactose) + glucose

The second Step: complex (β-galactosidase - galactose) + H_2O → galactose + β-galactosidase

In 1960, Wallenfels and Malhotra, based on many years of research speculated the hydrolysis mechanism on lactose, and believed that the active site β-galactosidase had two functional groups: thiol and imidazole, where thiol acts as a general acid and protonizes oxygen atoms of galactosidase, while imidazole may act as a nucleophile and attack the nucleophilic center of the first carbon atom of galactose molecule. When the galactosides receptor is water, hydrolysis occurs, and generates the product of glucose and galactose. In the course of the hydrolysis of lactose, β-galactosidase also has a certain function of transfer of glycosides: if the receptor is sugar or alcohol, transfer of galactosides occurs; if the receptor is lactose, it generates galacto-oligosaccharides with three glycosyls. The efficiency of transfer of glycosides of the enzyme depends on the source and nature of the enzyme, concentration of substrate lactose and reaction time [24].

The β-galactosidase from the fungal origin is extracellular, and can be produced by employing either solid-state fermentation or submerged fermentation for enzyme production, and production and extraction are easy. *A. oryzae* and *A. niger* are the primary strains for the production of β-galactosidase, and the enzymes produced by these fungi show acidic optimum pH, acid resistance, good chemical substances tolerance and good stability, and do not need stabilizer and activator. Therefore, the β-galactosidase from the fungal source is more suitable for the production of immobilized enzyme than that from yeast. The β-galactosidase from bacteria belongs to intracellular, and will not be secreted into the medium during the training, and possesses high heat resistance and is suitable for immobilization. The β-galactosidase from moderate or extreme *Thermcoccus litoralis* possesses the highest activity at temperatures of 70°C–90°C. In this temperature range, the immobilized enzyme can prevent the reaction system from bacteria contamination, and speed up the reaction rate, thereby enhancing the economic value of lactose hydrolysis. But due to the problems of galactotriose and toxicity produced by the bacteria, it has not been applied for industrial production. At present, the study of the β-galactosidase from *E. coli* is the most thorough: its enzymatic properties, structure of the enzyme protein and catalytic mechanism have been reported in detail; its active site is tyrosine residues and carboxyl site, which in biochemical analysis is widely used. The β-galactosidase from *B. circulans*, in the hydrolysis process, also has the activity of transfer of galactosides, and the galacto-oligosaccharides generated are mainly trisaccharides and tetrasacchrides.

β-galactosidase can be used to prepare galactosyl-CD. Sumio in Ensuiko Sugar Refining Co. Ltd, Japan, mentioned that a new type of galactosyl-α or β-CD could be prepared using β-galactosidase. They found that, using lactose as a donor and BCDs as a receptor, a variety of β-galactosidases from microoganisms could transfer the galactosyl to the side chain of BCDs, rather than directly connecting the CD ring. For the β-galactosidases from *Bacillus circulans*, *A. oryzae* and *Penicillium multicolor*, the structure of main galactosyl transfer product by them are different according to HPLC retention time in the synthesis of galactosyl branching CD [25].

The β-galactosidase from the *Pencillium simplicissimum* isolated from soil has high galactosyltransferase activity: Synthesis of galactooligosaccharides from 3 to 4 units of monosaccharide. When monosaccharides are present in the reaction system, the transfer activity of the enzyme will be inhibited, so only by the enzyme from *P. simplicissimum* cultured under the non-optimum hydrolysis conditions, galactooligosaccharide production can reach the maximum. For example, the lactose solution (concentration of 60%) incubated at pH 6.5 and 50°C obtained the maximum conversion rate. The product composition was 30.5% galacto-oligosaccharide, 27.5% residual lactose and 42% free monosaccharides.

2.8. Immobilization of the Enzymes for CD Preparation

Enzyme, as a biological catalyst, due to its catalytic efficiency, mild reaction conditions, high substrate specificity, environment-friendly features and easily controllable reaction, is widely used in food, textile, pharmaceutical and fine chemical industries. However, for a natural enzyme, the stability is poor: It is prone to denature in acids, alkalis, organic solvents and heat, resulting in the decrease or loss of its activity. It is easily mixed with reaction products, resulting in difficulty in purification, so it is difficult to carry out the enzyme reaction continuously and automatically, making it difficult to be more widely applied in industry. Therefore, we must apply enzyme engineering technology to promote further development of industrial enzymes, and immobilized enzyme is developed in this situation. In the CD production, the applications of immobilization technology are increasingly being used worldwide. The immobilized CGTase can be reused by simplifying the purification process of CD, which is of great significance to achieving continuous CD production. The immobilized approach of CGTase from basophilic *Bacillus no.*

38-2 was frequently studied in the past: it was acetylated first, and then adsorbed onto the vinyl pyridine copolymer. When this CGTase immobilization was applied to the non-continuous production of CD, no significant inactivation occurred after four times, while when applied to continuous column reaction for two weeks, 63% starch of its solution (4%) was transformed into the CD. Researchers continuously apply the new carrier and technology for immobilization of the CGTase. Professor Zhu Bo-Ru in Polymer Chemistry Institute of Nankai introduced chitosan into immobilization of CGTase, and investigated a series of its properties [26].

2.8.1. *Preparation of immobilized enzyme*

Immobilization of enzyme is the process where the free enzyme's catalytic activity is completely or largely restricted to a certain space, and is divided into embedded-type and bound-type. For embedded-type, enzyme itself (Fig. 2.22) does not bind, but is embedded in the monomer of the polymer, including plastic, microcapsule and liposome. In the case of bound-type, the enzyme is bound up to combine through the interaction between the enzyme and the carrier, and according to the different forms of combination, it can be divided into adsorption, covalent binding and crosslinking. In the immobilization of history of CGTase, including almost all of the above methods.

2.8.1.1. *Adsorption*

Adsorption is based on the interaction of secondary bonds on the surface of the carrier and enzyme. The first industrial-scale application of immobilized enzyme is the aminoacylase adsorbed to DEAE-Sephadex A25. Adsorption is the oldest, and easiest, economical immobilization method. According to the different characteristics of adsorption, it can be divided into: physical adsorption and ion exchange adsorption.

2.8.1.1.1. *Physical adsorption*

Physical adsorption is the method to use insoluble carrier, i.e., activated carbon, alumina, kaolin, bentonite, silica gel, porous glass, hydroxyapatite, cellulose, collodion, etc, and bind the enzyme to the insoluble carrier through a number of surface forces, i.e., hydrogen and hydrophobic bonds, etc. Physical adsorption is the most direct immobilization process and has several advantages: simple preparation, lower cost and many types of carriers to choose (some of them can be regenerated by treating with cleaning agents, strong denaturing agent or wetting agent). However, the interaction between the enzyme and the carrier is weak, and the adsorbed

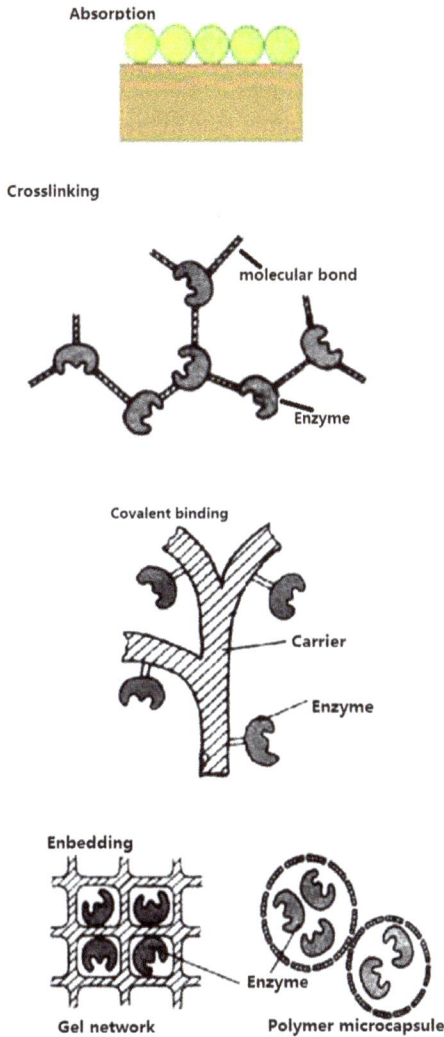

Fig. 2.22. The immobilization methods of CGTase [26].

enzyme is easy to fall off from the carrier. The non-specific adsorption of the enzyme can cause loss of some or all of the enzyme activity. Researchers have successfully used silica gel, diatomaceous earth, macroporous alumina, porous glass as carriers to immobilize a CGTase. The prepared immobilized CGTases have high efficiency, but sometimes are easy for desorption, so their span is usually limited. Keli *et al.* used commercial chitosan product to immobilize the CGTase produced from *B. firmus*. One approach was to use the physical adsorption method: first, chitosan was purified, and 10 mL of its solution was mixed with 2.0 mg/g of the CGTase solution, and then the mixture was incubated at a certain temperature and pH for immobilization. They found that, in the temperature range investigated (7°C–26°C), the fluctuations of temperature and mixing had little effect on the immobilized CGTase [27].

2.8.1.1.2. Ion exchange adsorption

For ion exchange adsorption, the enzyme is immobilized by the interaction between dissociation groups of the enzyme side chain and those of ion exchange under the appropriate pH and ionic strength conditions. The most common exchange agents are CM-cellulose, DEAE-cellulose, DEAE-Sephadex, polypolystyrene, Amberlite-97, Dowe X-50, etc. As mentioned earlier, the first industrial immobilized amino acid acylase was industrial produced by being adsorbed on the DEAE-cellulose and DEAE-Sephadex. In 1988, S. Sakai *et al.* described that the copolymers were obtained from glycidyl methacrylate and glycol methacrylate through copolymerization, and treated with triethylamine to obtain anion-exchange resin, and after treated with 1N NaOH, the resin could adsorb 68 mg/of CGTase, which had a good effect on the liquid starch (4%). Professor Zhu Bo-Ru, Nankai University, studied the immobilization of CGTase using spherical cellulose ion exchanger: to 1 mL carrier ball, 40 mL enzyme solution was added, and the mixture was incubated at pH 7.0 in phosphate buffer with shaking culture for certain time. Then, the enzyme solution was removed, and washed with pure water to protein-free out of the ball, and the immobilized CGTase was obtained. This enzyme showed that thermal stability and the *Km* increased. In the presence of 10 mmol/L $CaCl_2$, after storing at room temperature for 15 days, the enzyme retained 80% of the activity [28]. In another method of CGTase immobilization, ion-exchange paper or membrane was used as a carrier. For example, the immobilized CGTase NTE-370 prepared with ultrafiltration membranes containing amino can transform straight-chain dextrin into branching CD; the immobilized CGTase prepared employing another ultra-membrane containing amino acid can be used multiple times, and this would reduce production costs and improve efficiency; DEAE Zetap rep250 paper has also been used as ion-exchange carrier.

In recent years, a new method for preparing immobilized CGTase as emerged: 10 consecutive lysine residues are linked with enzyme, and then bound on the ion exchange agent. This method is based on the principle that polylysine helps protein to stabilize refold and purify. The characteristics of connected chemical bonds, pH effects, temperature stability and operational stability were investigated, and the results showed that the effect of polylysine residues on the noncovalent immobilization of CGTase is significant. Although the process of the immobilization of CGTase is reversible, in neutral and alkaline conditions, the interaction of the bonds between the immobilized complex is relatively stable.

2.8.1.1.3. The methods of preparation of the immobilized enzyme adsorption

i. Static method: Without stirring and shaking, the carrier is directly added to the enzyme solution, and through the processes of the natural adsorption, resolution and re-adsorption, the enzyme is immobilized, but this method is inefficient, time-consuming and unpredictable.

ii. Electrodeposition method: Two electrodes are placed in the enzyme solution, and the carrier is added to the region near the electrode and the electrodes are energized. The enzyme moves towards the electrodes and deposits on the surface of the carrier. Using this method, carrier and enzymes need to determine in advance whether it will break down and damage in the electric field.

iii. Dynamics batch immobilization method, also known as mixing, and shaking loading method: This is a commonly used laboratory method, and as compared to the static method, after carrier is mixed with enzyme solution, the immobilization processes are completed under continuous shaking using stirring or shaker. By employing appropriate stirring rate or shaking speed, the structure of the enzyme and the carrier are not destroyed, and the enzyme and carrier can be mixed fully to achieve an even fixed purpose.

iv. Reactor serial method: This is a commonly used industrial method and its characteristics are a combination of the immobilization and subsequent application, that is, first, carrier is mounted in the reactor, and the enzyme solution is added. Then mix fully the enzyme with the carrier through vibration modes of circular flow.

2.8.1.2. *Covalent binding method*

Covalent binding method, also known as covalent coupling method, is the method of immobilization of enzyme by the coupling combination of the groups of side-chain that are non-necessary to the activity of the enzyme protein molecules and the functional groups of carriers. This approach is the most common method in the carrier binding method that has been reported, and is the most actively large class of methods in immobilized enzyme studies. The advantage of this approach is that the coupling interaction between the enzyme and the carrier is very strong, and the enzyme is not lost in the course of usage and has good stability. The disadvantage is that the operations of carrier activation process are more complicated, and the reaction conditions are relatively severe. In addition, the structure of the target enzyme may be changed by participating in enzyme reactions, leading to a decline in activity. Therefore, in the immobilization process, in order to obtain products with a higher activity, reaction conditions must be strictly controlled.

Characteristics of the carrier itself are directly related to the formation and the nature of the immobilized enzyme. In general, hydrophilic carriers are better than the hydrophobic carrier in the amount of protein bound, the viability and stability of the immobilized enzymes. Ideal carriers should have the following characteristics: (1) loose structure, large surface area, and moderate pore size; (2) have a certain mechanical strength, the coupling group that match side chain groups of the target enzyme, (3) little or no non-specific adsorption. Common carriers include: natural polymers (cellulose, agarose, starch, glucose gel, collagen and its derivatives, etc), synthetic polymers (nylon, poly-amino acid, ethylene-maleic anhydride copolymer, etc), inorganic support materials (porous glass, metal oxides, etc). Zaita *et al.* covalently bound extracellular endo-inulinase obtained from *A. niger* to agarose (CAS), activated by bromoacetyl cellulose (BAC), cellulose carbonate (CC) and cyanogen bromide, and stimulated the three immobilized enzymes using the acidic chloride of mercury, Fe^{3+} and Mn^{2+} to improve their abilities to resist inactivation. Results showed that when the CAS immobilized enzyme was applied to hydrolyze inulin, the optimum temperature significantly increased (from 45°C up to 60°C); the stability of BAC immobilized endo-inulinase decreased in the acidic environment, and increased in an alkaline environment; the stability of CC immobilized endo-inulinase was relatively high in the acidic environment. The Km of BAC immobilized endo-inulinase decreased, while that of the CAS immobilized endo-inulinase increased [29].

For the covalent bond method, activation of carrier and the operations are relatively complex, and the conditions must be strictly controlled to increase the immobilized enzyme activity. The group types that participate in covalent linking to the protein, and the physical and chemical properties of carrier have an impact on the immobilized enzyme. Therefore, it is often combined with cross-linking and adsorption. Keli *et al.* used commercial chitosan product to immobilize CGTase produced by *B. firmus*: the chitosan carrier was mixed with a certain amount of hexamethylene diamine, or γ-aminopropyl triethoxysilane (APTES), resulting in covalent bonding to each other. 3% glutaraldehyde was added to the carrier mixture according to certain rate, which mainly acted as a crosslinking agent, and after filtering, washing and vacuum drying, etc, the CGTase was immobilized by adsorption [27]. This method is also known as Schiff alkaline: the carrier chitosan with primary amines is activated with glutaraldehyde, and the remaining aldehydes react with the primary amines in the enzyme, resulting in immobilizing the enzyme. The carrier along with aldehydes such as Silochrame C-80 can also directly immobilize the CGTase. In order to overcome the steric hindrance, Cui *et al.* in the study of CGTase immobilization, also used diamine as spacer, resulting in improvement in the immobilized enzyme activity. Beaded agarose gel and chitosan were activated by halogenated cyanide, e.g., CNBr, and under alkaline conditions, coupled enzyme, which was also reported to be used for the immobilization of CGTase. T. Nakakuki immobilized CGTase using this method, which retained 70% of the activity in use after 30 days [30].

2.8.1.3. *Crosslinking*

Similar to the covalent bonding method above, crosslinking is carried out through chemical combination. The difference is that crosslinking has its own characteristics. It is the crosslinking between the enzyme molecule itself, or between the enzyme molecules and the inert protein, or between the enzyme molecules and the carrier using multifunctional reagents. It is the way to immobilize the enzyme molecule through the formation of covalent crosslinked frame structure. Commonly used crosslinking reagents are glutaraldehyde, dual nitrogen benzidine 2,2 '- disulfonic acid, sorbic acid azomethine acid ester and toluene-2-cyanuric-4-different thiocyanate, etc, of which the most widely used is glutaraldehyde. Jyoti L Iyer found that although using purified sand to fixed CGTase achieved an absorption rate of 98%, the enzyme was easy to fall off from the media, and increased bonding by glutaraldehyde crosslinking and solved this problem. After eight batch enzyme reaction, it was still able to maintain 80% of the activity. Rohrbach adsorbed the CGTase to the alumina surface, and then crosslinked with glutaraldehyde. The conversion rate using 2% starch solution by the immobilized enzyme reached 39% at 50°C. Rozzell used silica gel as a carrier and glutaraldehyde as the crosslinking agent to immobilize CGTase, and the immobilized enzyme activity and stability improved greatly as compared with the free enzyme. Ivony *et al.* immobilized CGTase using the polyacrylamide as carrier and water-soluble carbodiimide as crosslinker, resulting in decent immobilization.

2.8.1.4. *Embedding*

The monomer of the polymer is mixed with the enzyme solution to aggregate with the aid of polymerization accelerator (including the crosslinking agent), and thus the enzyme is embedded in the polymer to achieve immobilization. In terms of the target enzyme, it is primarily a physical embedding (not chemically modifying enzyme amino acid residues), so the method will rarely change the structure of the enzyme, and is generally more secure. However, in the chemical polymerization process, the reaction between other substances in system will be accompanied by free radical generation and heat production. Chemical reactions may occur between enzyme and reagents often leading to enzyme inactivation. Therefore, in the design of embedding conditions, these factors must be considered. Guo *et al.* added the solution of CGTase produced by *Bacillus alcalophilus* to the appropriate amount of sodium alginate which was then fully mixed, and placed at room temperature for 1 h. Then, they drew the mixture into the $CaCl_2$ solution using sterilized syringes to make gel beads, placed gel beads at room temperature for a period of time to harden, and obtained hardened gel beads by leaching, and then removed $CaCl_2$ by washing with distilled water, and finally got the immobilized CGTase. The optimal conditions of the sodium alginate-embedded CGTase were determined through experiments and were as follows: sodium alginate concentration 7% (final concentration about 2.3%), $CaCl_2$ concentration 3%, ratio of sodium alginate to the enzyme solution volume 1:2 and immobilization time of 2.5 h. The properties of the immobilized enzymes, including optimum temperature and optimum pH, did not change significantly as compared with the free enzyme, while the thermal stability and pH stability were significantly higher than those of free enzyme, and the immobilized CGTase retained 75% of enzyme activity for seven consecutive batches [31].

2.8.2. *Properties and application of immobilized enzyme*

Immobilization may have some impact on enzymes and their surroundings. Immobilized enzyme will show different characteristics as compared to the enzyme solution. As to most of the enzymes, after the immobilization, their stability, longevity and shelf life will greatly increase, but their own activities show a certain decline, and the catalytic characteristics of the free enzyme are also different. When recognizing and mastering these rules of the immobilized enzyme, we can effectively guide the usage.

2.8.2.1. *Immobilized enzyme activity*

Compared with the activity of the enzyme solution, the immobilized enzyme activity is generally declined. The reasons are many: (1) The combination of the immobilized enzyme with insoluble carrier causes changes in the structure; (2) The combination of the important amino acid residues of the enzyme active site and carrier changes the conformation of enzyme active site or adjustment of the center; (3) The interaction between enzyme–substrate is subject to the steric hindrance, substrate and effector of enzymes which cannot directly contact, thus affecting the activity causing the hydrophilic, hydrophobic and dielectric constants of the immobilized carrier to often interfere with the activity.

2.8.2.2. *The catalytic properties of immobilized enzyme*

The characteristics of immobilized enzymes, such as substrate specificity, optimum pH of enzyme reaction, *Km*, the maximum reaction rate and optimum temperature, etc, will differ from those of the free enzyme, because of the different immobilization processes.

2.8.2.2.1. Substrate specificity

The substrate specificity of an immobilized enzyme is usually consistent with that of the enzyme solution, but when an enzyme is immobilized by a water-insoluble carrier, due to the steric effects caused by steric hindrance, for the high molecular weight substrate, the substrate activity of the immobilized enzyme will be greatly reduced. For CM-cellulose azide derivatives immobilized glucoamylase, the molecular weight of 8,000 amylose activity is 77% of that of free enzyme, while the relative molecular weight of 500,000 amylose activity is only 15%–17% of that of free enzyme. As to the immobilized CGTase, its substrate specificity does not significantly change, and the substrates are still mainly starch and malt polysaccharides. When soluble starch, corn starch, potato starch, rice starch, amylose and amylopectin are used as substrates, the activities of free enzyme and immobilized CGTase are not very different, while when dextrin and the half liquefied starch are used as substrates, the immobilized enzyme activity increased about 1.2 to 1.3 times as compared with the free enzyme because small molecules can more easily contact with the immobilized CGTase.

2.8.2.2.2. The optimum pH of enzyme

Compared with free enzyme, the substrate optimum pH value and pH curve of immobilized enzyme are often shifted because of the charge of the enzyme protein or water-insoluble carrier: in general, for the immobilized enzyme prepared by anionic polymer carrier, due to the attraction effect of the carrier on cations including H^+ in the solution. H^+ concentration around the immobilized enzyme will increase, so the external pH of the solution must be alkaline in order to offset the environmental effects for the greatest activity. On the contrary, the

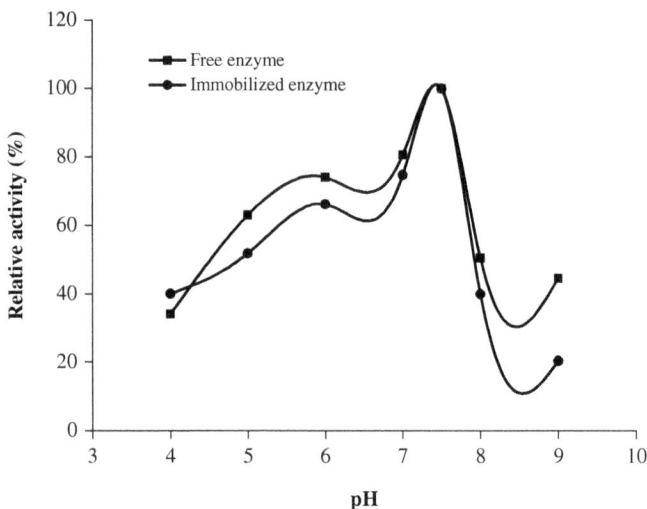

Fig. 2.23.   Effect of pH on the CGTase from *Bacillus macerans* ATCC 8244 and immobilized CGTase [32].

enzyme immobilized by the carrier with positive charge has acidic optimum pH. However, pH will not be offset for all the immobilized enzymes. Arya and Srivastava immobilized CGTase from *B. macerans* ATCC 8244 using alginate, at a temperature of 60°C and pH range of 4–9, and investigated the stability of the free enzyme and immobilized enzyme (Fig. 2.23). The results showed that: (1) The optimum pH of the free CGTase and immobilized CGTase are the same 7.5; (2) in the range of pH 4–7, the immobilized enzyme activity was higher than that of the free enzyme by about 30%–80%, but at a higher pH value, the activity decreased by 40% [32].

2.8.2.2.3. The optimum temperature or the immobilized enzyme

Enzyme activity has a certain dependence on temperature, and the general enzyme has an optimal temperature, and immobilized enzyme is no exception. Compared to the solution enzyme, the optimum temperature of immobilized enzyme shows ups and downs. The study found that the aminoacylase was bound to DEAE-cellulose and DEAE-dextran using ion-binding, or embedded in crosslinked polyacrylamide gel. Thus, the immobilized enzyme was prepared, and its optimum temperature was somewhat higher than that before the immobilization. When aminoacylase was immobilized by iodine acetyl cellulose with covalent binding, its optimum temperature was somewhat lower than that before the immobilization. When the glucose isomerase was bound to porous resin with covalent binding,

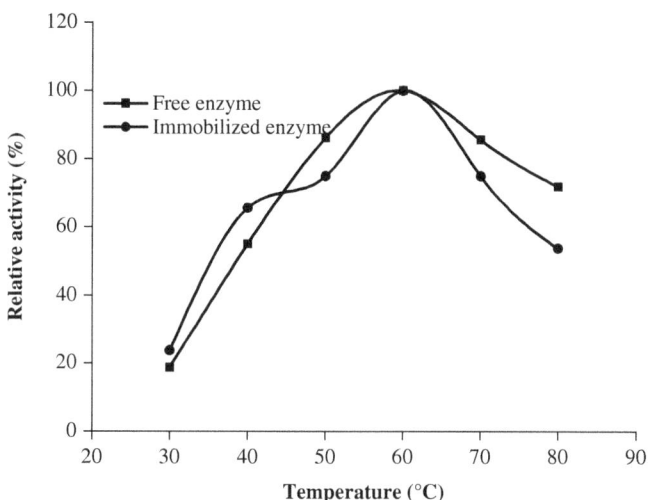

Fig. 2.24. Effect of temperature on the free and immobilized CGTase from *Bacillus macerans* ATCC 8244 [32].

and when converting enzyme was fixed on bentonite with physical adsorption, or embedded in polyacrylamide gel, the optimum temperatures for the immobilized enzymes were identical to those for the free enzymes. Arya and Srivastava immobilized the CGTase from *Bacillus macerans* ATCC 8244 using alginate, and investigated the changes in the activity of the free and immobilized CGTase in the temperature range of 30°C–80°C [32] (Fig. 2.24). The activities of free and immobilized CGTase increased with the increasing of temperature, and when the temperature reached 60°C, the activity was the highest. This shows that the effect of temperature on the activity of free CGTase was the same as that of immobilized CGTase, and did not result in greater loss of enzyme activity.

2.8.2.2.4. Michaelis constant (*Km*) and maximum reaction rate

Km reflects the affinity between the enzyme and the substrate, and the *Km* of immobilized enzyme changes little or much, depending on the interaction between immobilized enzyme and carrier. When enzyme is immobilized using carrier binding, due to the electrostatic interaction between the immobilized enzyme and the carrier, *Km* of the immobilized enzyme decreases. Maximum reaction rate may differ in terms of fixed methods. The maximum reaction rate of the invertase, immobilized by porous glass using covalent binding method is the same as the free enzyme; while the maximum reaction rate of the invertase embedded by

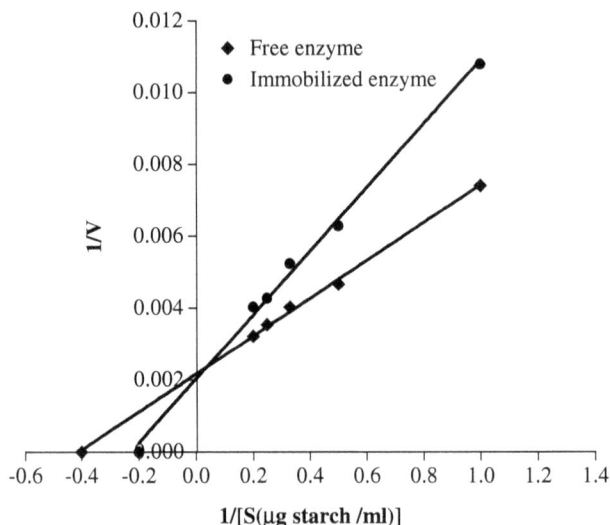

Fig. 2.25.   The Lineweaver Burk diagram of alginate immobilized CGTase [32].

polyacrylamide gel is 10% lower than that of the free enzyme. Arya and Srivastava immobilized the CGTase from *B. macerans* ATCC 8244 using alginate, and by Lineweaver Burk plot method, they found that the *Km* of immobilized CGTase increased from 2.5 to 4.5 μg/mL starch (Fig. 2.25), which was consistent with most results of the immobilized CGTase system studied [32]. The *V*max of the free CGTase was 475 μg β-CD/(mL·min), while the *V*max of the immobilized enzyme was 515 μg β-CD/(mL·min). After a period of reaction time, due to the increasing product concentration, CGTase activity began to decrease. The constant of the decomposition (KI) of the complex formed can be obtained by the formula: $Smax^2 = Km \times KI$, where *Smax* is the concentration of the substrate matrix. The KI of the free and immobilized CGTase were 10 μg starch/mL and 5.5 μg starch/mL, respectively, indicated that the immobilized CGTase showed lower substrate sensitivity as compared to the free enzyme.

2.8.2.3. *Immobilized enzyme stability*

After immobilization, the stability of the enzyme toward various reagents, protein-degrading enzyme, high temperature, storage at low temperature and operation increases to some extent. However, the heat inactivation of immobilized enzyme will occur in varying degrees, and the catalytic activity will gradually decrease. From an economic point of view, the life of the immobilized enzyme is an extremely important parameter: the longer life expectancy, the longer is the continuous operation of the industry and the lower is the cost of production.

2.8.2.3.1. Thermal stability

Most of the heat resistance of the immobilized enzyme has improved to some extent, and the thermal stability of some immobilized enzymes increase up to 10 times due to immobilization, however, the mechanism is difficult to describe clearly.

2.8.2.3.2. The stability toward various reagents

The stability toward various reagents is listed in Table 2.12: After immobilization, the resistance of most of the enzymes increases toward protein denaturing agents or inhibitors. Even more interesting aspect is that some enzymes such as aminoacylase and trypsin will not only be not affected by the protein denaturant such as urea, guanidine hydrochloride, etc, but in the presence of these denaturants, the enzyme activity would be enhanced. This phenomenon may be attributed to the increased enzyme flexibility by these denaturants. However, as seen in the case of glucoamylase in Table 2.12, after immobilization, they become more sensitive to certain inhibitors. Effect of surfactants and metal ions on immobilized CGTase activity is shown in Table 2.12: the activities of immobilized CGTases are free from

Table 2.12. Effect of surfactant and metal ions on the activity of the immobilized CGTase [32].

Materials	Concentration	Relative activity of free enzyme (%)	Relative activity of immobilized enzyme (%)
—	—	100	100
Twain-80	0.01%	89	100
Twain-40	0.01%	83	98
Trinitrotoluene-100	0.01%	92	100
Sodium dodecyl sulfate	0.1%	44	91
$CaCl_2$	10 mmol/L	100	100
$CuSO_4$	10 mmol/L	67	87
$FeSO_4$	10 mmol/L	44	79
$HgCl_2$	10 mmol/L	41	68
$CoCl_2$	10 mmol/L	88	94

the surfactants and Ca^{2+}, but Cu^{2+}, Fe^{2+}, Hg^{2+} and Co^{2+} inhibit the enzyme activity. Since in the immobilized enzyme production process, the carrier will change the molecular structure of the enzyme, making it more difficult for the metal ion to be close to the enzyme active site. Thus the impact on the immobilized enzyme is less than that on the free enzyme.

2.8.2.3.3. Stability toward protease

After immobilization, the resistance toward protease increases. The reason may be an increase of steric access hindrance to the immobilized protease due to the three-dimensional obstacles of the carrier. If we take aminoacylase for example, under the action of trypsin, the activity of the solution enzyme was only 23%, the activity of the enzyme immobilized by DEAE-cellulose was 33% and the activity of the enzyme immobilized by DEAE-dextran was 87%. Under the action of protease, Pronase P, the activity of the solution enzyme, the enzyme immobilized by DEAE-cellulose and the enzyme immobilized by DEAE-dextran was 48%, 53%, 88%, respectively. Immobilized protease can generally avoid digestion damaging itself.

2.8.2.3.4. Operational stability

In a continuous reaction, enzyme stability is an important factor to determine whether the immobilized enzyme can be applied to industrial production. Only

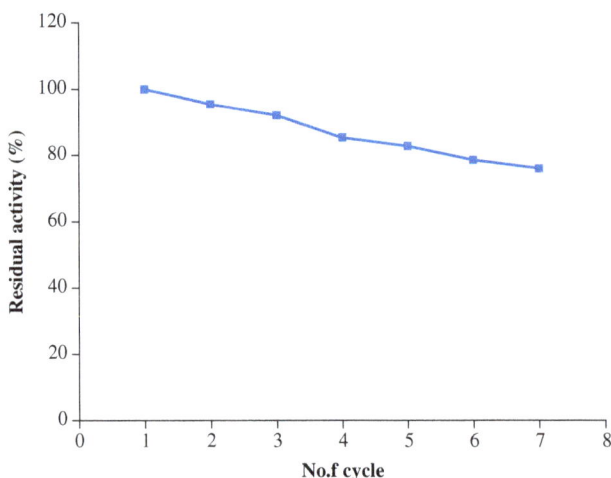

Fig. 2.26. The operational stability of immobilized CGTase immobilized by sodium alginate [32].

the immobilized enzyme with half-life of more than one month has industrial application value. The CGTase from *B. macerans* ATCC 8244 was immobilized by Arya and Srivastava using sodium alginate. After continuous fermentation for seven days (Fig. 2.26), the immobilized CGTase retained 75% activity [32]. This also shows that the loss of enzyme activity may be mainly a physical loss, and the enzyme itself changes little in nature. As the longer half-life of this enzyme is obtained, immobilized enzyme can be widely applied in the food industry.

2.8.2.3.5. Storage stability

The storage stability of most immobilized enzyme is greater than that of the free enzyme. The studies on the immobilized CGTase showed that, due to the new bonds between the enzyme and the carriers, the resistance of the immobilized CGTase on the environmental factors increases.

References

1. Tonkova, A (1998). Bacterial cyclodextrin glucanotransferase. *Enzyme and Microbial Technology*, 22, 678–686.
2. Van Der Veen, BA, JC Uitedehaag, BW Dijkstra and L Dijkhuizen (2000). Engineering of cyclodextrin glycosyltransferase reaction and product specificity. *Biochimica et Biophysica Acta*, 1543, 336–360.
3. Machovic, M and S Janecek (2007). Amylolytic enzymes: Types, structures and specificities. *Industrial Enzymes*, 3–18.
4. Rimphanitchayakit, V and Y Sakano (2005). Construction of chimeric cyclodextrin glucanotransferases from *Bacillus circulans* A11 and Paenibacillus macerans IAM1243 and analysis of their product specificity. *Carbohydrate Research*, 340(14), 2279–2289.
5. Davis, GJ, KS Wilson and GE Henrissat (1997). Nomenclature for sugarbinding subsites in glycosyl hydrolases. *Biochemical Journal*, 321, 557–559.
6. Uitdehaag, JC, KH Kalk, BA Veen, *et al.* (1999). The cyclyzation mechanism of cyclodextrin glucanotransferases (CGTase) as revealed by a gamma- cyclodextrin-CGTase complex at 1.8-A resolution. *The Journal of Biological Chemistry*, 274, 34868–34876.
7. Fujita, Y, H Tsubouchi, Y Inagi, K Tomita, A Ozaki and K Nakanishi (1990). Purification and properties of cyclodextrin glycosyltransferase from Bacillus sp AL-6. *Journal of Fermentation and Bioengineering*, 70, 150–154.
8. Sorman, BE and ST Jorgensen (1992). Thermoanerobacter SP. CGTase: Its properties and application. *Denpun Kagaku*, 39, 101–108.
9. Nomoto, M, C Cheng, C Dey and S Sheu (1986). Purification and Characterization of Cyclodextrin Glucanotransferase from an alkalophilic bacterium of Taiwan. *Agricultural and Biological Chemistry*, 50(11), 2701–2707.
10. Zhang, QB, ZZ Yan and SZ Zhang (1998). Purification and properties of cyclodextrin glucosyltransferase from alkaliphilic bacteria. *Acta Microbiologica Sinica*, 38(2), 98–102. (in Chinese)

11. Abelian, VA, MO Adamian, LA Abelian, AM Balayan and EK Afrikian (1995). A new cyclomaltodextrin glucanotransferase from *Bacillus halophilic. Biokhimiya*, 60, 891–897. (in Russian)
12. Cao, XZ and ZY Jin (2005). Optimization study on cyclodextrin glycosyltransferase production by *Bacillus alkalophilus. Food Science*, 26(2), 122–126. (in Chinese)
13. Rosso, A, S Ferrarotti, MV Miranda, N Krymkiewicz, BC Nudel and O Cascone (2005). Rapid affinity purification processes for cyclodextrin glycosyltransferase from *Bacillus circulans. Biotechnology Letter*, 27, 1171–1175.
14. Kunamneni, A, T Prabhakar, B Jyothi and P Ellaiah (2007). Investigation of continuous cyclodextrin glucanotransferase production by the alginate-immobilized cells of alkalophilic *Bacillus sp.* in an airlift reactor. *Enzyme and Microbial Technology*, 40, 1538–1542.
15. Nakamura, N and K Horikoshi (1976). Purification and properties of neutral-cyclodextrin glycosyl-transferase of an alkalophilic *Bacillus sp. Agricultural and Biological Chemistry*, 40, 1785–1791.
16. Rha, CS, DH Lee, SG Kim, WK Min, SG Byun, DH Kweon, NS Han and JH Seo (2005). Production of cyclodextrin by poly-lysine fused *Bacillus macerans* cyclodextrin glycosyltransferase immobilized on cation exchanger. *Journal of Molecular Catalyci B: Enzymatic*, 34, 39–43.
17. Kitahate, S, S Ishikawa, T Miyata and O Tanaka (1989). Production of rubusoside derivatives by transglycosylation of various beta-galactosidase. *Agricultural and Biological Chemistry*, 53, 2923–2928.
18. Tang, BY, XH Zhu and J Liu (2001). Selection of the acid and heat-resistant pullulanase-producing strain and its fermentation conditions. *Microbiology*, 28(1), 39–43. (in Chinese)
19. Jin, Y, GX Gu, Lu and J (1999). Effect of medium and culture conditions on pullulan. *Journal of Wuxi University of Light Industry*, 18(2), 33–38. (in Chinese)
20. Ma, XJ, XJ Zhang and R Wang (2002). Studies on screening and breeding of alkaline pullulanase producing bacterium and regulation of nutrition, *Acta Botanccu Boreal*, 22(4), 883–888. (in Chinese)
21. Kubo, A, N Fujita, K Harada, T Matsuda, H Satoh and Y Nakamura (1999). The starch-debranching enzymes isoamylase and pullulanase are both involved in amylopectin biosynthesis in rice endosperm. *Plant Physiology*, 121, 399–409.
22. Sugimoto, T, A Amemura and T Harada (1974). Formations of extracellular isoamylase and intracellular alpha-glucosidase and amylase(s) by Pseudomonas SB15 and a mutant strain. *Applied Microbiology*, 28, 336–339.
23. Shen, WY (2009). Preparation, properties and application of α-galactosidase from 33 germinating coffee beans. Ph.D thesis. Wuxi: Jiangnan University. (in Chinese)
24. Wallenfels, K and OP Malhotra (1960). In PD Boyer, H Lardy and K Myrbäck The Enzymes (eds.), 2nd edn., vol. 4, pp. 409–430 New York, Academic Press.
25. Sumio, K (1989). Production of rubusoside derivatives by transgalactosylation of various β-galactosidase. *Agricultural and Biological Chemistry*, 53(11), 2923–2928. (in Japanese)
26. Zhu, BR, ZQ Shi and BL He (1996). The immobilization of cyclodextrin glycosyltransferase and its usage: (II) study on immobilization of cyclodextrin glycosyltransferase

using chitosan as support. *Ion Exchange and Adsorption*, 12(6), 532–536. (in Chinese)

27. Keli, CAS, MORD Regina, DEO Rogerio, D Oliveira, FFD Moraes and GM Zanin (2002). Immobilization of cyclodextrin glycosyltransferase (CGTase) from *Bacillus firmus* in commercial chitosan. *Journal of Inclusion Phenomina and Macrocyclic Chemistry*, 44, 383–386.

28. Zhu, BR, ZQ Shi and BL He (1997). The immobilization of cyclodextrin glycosyltransferase and its usage: (I). Study on immobilization of cyclodextrin glycosyltransferase using anion cellulosic beads as support. *Acta Scientiarum Naturalium Universitatis Nankaiensis* (Natural Science Edition), 30(1), 88–93. (in Chinese)

29. Zaita, N, T Fukushige, M Tokuda, K Ohta and T Nakamura (2000). Preparation and enzymatic properties of *Aspergillus niger* endoinulinase immobilized onto various polysaccharide supports. *Food Science and Technology Research*, 6, 34–39.

30. Cui, HJ, ZM He and CY Jia (1993). Study on immobilization of cyclodextrin glycosyltransferase using Konjac glucomannan. *Natural Product Research and Development*, 5(3), 48–54. (in Chinese)

31. Guo, Y, WW Wang, LW Guo and LM Xiao (2007). Immobilization of cyclodextrin glycosyl transferase in sodium alginate. *Food and Fermentation Industries*, 33(9), 33–36. (in Chinese)

32. Arya, SK and SK Srivastava (2006). Kinetics of immobilized cyclodextrin glucanotransferase produced by *Bacillus macerans* ATCC 8244. *Enzyme and Microbial Technology*, 39, 507–510.

3

PREPARATION AND ANALYSIS OF CYCLODEXTRIN

An-Wei Cheng*, Jin-Peng Wang† and Zheng-Yu Jin†

*Institute of Agri-food Science and Technology,
Shandong Academy of Agricultural Science,
Jinan 250100, China
†The State Key Laboratory of Food Science and Technology,
School of Food Science and Technology, Jiangnan University,
Wuxi 214122, China

As far as we know, almost all kinds of cyclodextrins (CDs) are produced by enzymatic techniques, in which cyclodextrin glucosyl transferase enzyme (CGTase) has been widely used for industrial production of common α-, β- and γ-CDs. Only a few reports have focused on the chemical synthesis processes: A CD with a degree of polymerization DP of five (cyclomaltopentaose) was chemically synthesized by Nakagawa *et al.* in 1994 [1]; CD6 and CD9 were chemically synthesized by Wakao *et al.* [2] through tethering the donor to the acceptor by the phythaloyl bridge, using maltose or maltotriose as a precursor. The total processes comprised 26 steps [2]. The chemical synthesis processes are too complicated to be used in industrial production. Only processes of enzymatic synthesis of CDs (α-, β-, γ-CD and large-ring CD) are described in this chapter.

3.1. Introduction

3.1.1. *Enzymatic preparation of cyclodextrins*

During the process of enzymatic conversion of starch, the organic solvent is usually added in the reaction system to form complexes with CDs. The complex can precipitate and thus has to be separated from the reaction system. However, the added organic solvents have potential safety risks, and would cause environmental contamination. Thus, a non-organic solvent system for CD preparation has also been developed. CD preparation processes are divided into non-control system and control system, depending on whether the organic complex additives have been added in the reaction system or not.

3.1.1.1. *Non-control system (add non-organic solvent)*

The CD preparation process without organic solvent is carried out based on the different physical properties of the CDs, such as solubility, molecular size, etc. The non-control system for CD preparation is as follows: After several hours' of reaction between CGTase and starch, residual insoluble starch and dextrin are removed from the reaction solution by centrifugation, and the filtrate is then decolored and desalted. For β-CD preparation, the refined filtrate is concentrated and kept under low temperature, then β-CD can crystallize because of its low solubility, and high purity β-CD can be obtained by recrystallization. For getting α-CD, the initial procedure is the same as that of β-CD, but for the removal of β-CD first, then add α-amylase into the filtrate to degrade residual soluble dextrin, concentrate the solution and keep it at low temperature. As a result, 50%–70% of α-CD would be crystallized. For γ-CD preparation and separation, gel chromatography is commonly selected because of its high solubility and poor crystallization property. Finally, 40%–60% of γ-CD could be obtained by repeating the procedure.

The byproducts, i.e., monosaccharide and disaccharides, formed during the reaction process are inhibitors for CD production. Removing these saccharides by ultra-filtration membrane system can significantly improve the yield of the CDs. Furthermore, the membrane can also prevent the loss of CGTase, which would make the reaction process continuous and efficient. An equipment of non-fixed ultra-filtration membrane combined with reverse osmosis membrane has been successfully developed for CD production [3]. In brief, at a low substrate concentration, the reaction solution was transferred to the ultra-filtration system during cyclization by CGTase, and the products were concentrated by the reverse osmosis membrane.

3.1.1.2. *Control system (add organic solvent)*

The preparation of CDs is carried out by adding organic solvents that are based on the different inclusion properties of CDs. Cyclohexane and decanol are suitable for α-CD preparation, toluene is suitable for β-CD preparation and organic compounds while more than 12 rings are suitable for γ-CD preparation. In addition, the CD yields would be changed by adding organic solvents even in the same reaction system. For example, when cyclododecanones are added as complex compounds into the reaction system of *Bacillus macerans* CGTase and 10% potato starch, the yields of β-CD and γ-CD are 98.5% and 1.5%, respectively, whereas, using cycotridecanones as complex compounds, the yields of the two CDs are 2% and 98%, respectively [4].

3.1.2. *Influence factors for CDs preparation*

3.1.2.1. *Enzyme source*

CGTase from different strains exhibits different relative specificity. The specificity for α-CGTase is relatively worse, though α-CD is the main product at the beginning of the reaction. The concentration of β-CD exceeds that of α-CD after one hour's reaction. Specificity for β-CGTase is relatively good and the dominant product of the β-CGTase reaction system is β-CD. Few γ-CGTases are found in naturally occuring strains, but the strain mutation would result in special reaction of γ-CGTase to starch. Cao *et al.* reported that *Bacillus alkalophilus* 1177 was mutated by γ-ray and produced γ-CGTase with the ability of producing γ-CD [5]. The enzyme source for α-, β- and γ-CGTases are shown in Table 3.1.

3.1.2.2. *Reaction substrate*

The starch materials (including corn, potato, cassava, soluble starch and waxy starch) were commonly used as substrates when producing CDs. Under the same

Table 3.1. CGTases producing CDs.

Enzyme	Strain source	Main product	References
CGTase (EC2.4.1.19)	*Klebsiella pneumoniae* M 5al	CD6	7
	B. macerans IAM 1243	CD6	8
	K. pneumoniae AS-22	CD6	9
	Alcalophilic Bacillus sp. strain A2-5a	CD7	10,11
	B. macarans	CD7	10–12
	Bacillus stearathermophilus	CD7	10
	Bacillus circulans	CD7	11
	Paenibacillus sp. F8	CD7	11
	B. alkalophilus 1177	CD8	5
	Bacillus subtilis no. 313	CD8	13
	Alkalophilic Bacillus strain 290-3	CD8	14
	Bacillus sp. AL-6	CD8	15

reaction conditions, the varied substrate resulted in different yields of CDs. Cao selected maize, cassava, canna edulis and potato starch as substrates, and showed that cassava and canna edulis starches resulted in much higher yields of γ-CD [5]. Zhang *et al.* found that corn starch was the appropriate substrate compared with potato and sweet potato starch for CGTase from *Bacillus alcalophilus* NK23 [6]. In the food industry, the widely available corn starch and cassava starch are commonly used as substrates to produce CDs.

The substrate concentration has a significant effect on the yield of CDs. Taking CGTase from *Alkalophilic Bacillus* as an example [5], the yields of α-CD, β-CD and γ-CD are 6.9%, 43.8% and 23.4%, respectively, when using 2.5% canna edulis starch as a substrate. The yields of α-CD, β-CD and γ-CD are 25.6%, 18.5% and 16.7%, respectively when the substrate concentration is 7.5%. While for 10% substrate, the yields of α-CD, β-CD and γ-CD are 15.8%, 14.2%, 13.9%, respectively. Very high concentrations of starch would result in reduction of CD yields. Considering the economic efficiency, 30% starch was commonly used in industrial production.

3.1.2.3. *pH value*

The characters of CGTase from varied strains are different. For example, the optimal pH of CGTase from *B. circulans* C31 is 5.5, the stabled pH varied from 5.5 to 9.0 [16]; the optimal pH of CGTase from *B. macerans* is 5.0–5.7, the stabled pH varied from 5.0 to 10.0 [17]; the optimal pH of CGTase from *Alkalophilic Bacillus* is 8.0, the stabled pH varied from 8.0 to 10.0 [18].

3.1.2.4. *Temperature*

The optimal and stable temperatures for various CGTases are different. For example, the optimal temperature and stable temperature for CGTase from *B. circulans* C31 are 60°C and 50°C, respectively [16]; the optimal and stable temperatures are 60°C and 70°C, respectively for CGTase from *B. macerans* [17].

CGTase exhibits four kinds of transglycosylation reactions including coupling, disproportionation, hydrolysis and cyclization, and these four reactions are influenced by the reaction temperature. Qi *et al.* compared the effects of different reaction temperatures on the cyclization and coupling reactions of the CGTase from *B. macerans* influencing the yield of large ring CDs (LR-CD) synthesis products. They found that higher yields of LR-CD were obtained with a reaction temperature of 60°C compared to 40°C [18]. Endo *et al.* also found that the composition of LR-CD was greatly affected by the reaction temperature [19].

3.1.2.5. *Reaction time*

All reactions need some time to reach equilibrium, thus, reaction time is important for the CD yield and conversion rate. Commonly, the reaction time for β-CD production is no more than 30 h in the industry. The reaction time of obtaining the maximum yield of α-, β- and γ-CDs is different even if the substrate and enzyme is the same.

3.1.2.6. *Organic solvent addition*

The yield of CDs could be enhanced by adding appropriate organic solvent in the system. Chen *et al.* reported that the yield of CDs increased about 100% when adding 10% of ethanol [20]. Several water-insoluble organic solvents such as decane and toluene can also be used to increase the yield of CDs as described in Sec. 3.1.1.2. However, those solvents are prohibited for CD production when CDs are used in food because of the toxicity.

3.2. Industrial Process for α-Cyclodextrin Preparation

3.2.1. *Control system*

The process of α-CD preparation in a control system is shown in Fig. 3.1.

Following the procedure described in Fig. 3.1., final α-CD with the purity of >99% can be obtained.

3.2.2. *Non-control system*

The process of α-CD preparation without organic solvent is shown in Fig. 3.2.

Following the procedure described in Fig. 3.2., α-CD with purity of >99% can be obtained.

3.3. Industrial Process for β-Cyclodextrin Preparation

3.3.1. *Control system*

The process of β-CD preparation in control system is shown in Fig. 3.3.

Following the procedure described in Fig. 3.3, β-CD is obtained with the purity of 99.7%.

3.3.2. *Non-control system*

The process of β-CD preparation without organic solvent added is shown in Fig. 3.4.

Liquefied starch

↓CGTase from *cerans*

Enzymatic conversion at 60°C

↓

Unconverted saccharides are hydrolyzed by glucoamylase

↓

Concentration

↓

Precipitation with ethanol, filtration

↓

Decolored with active carbon

↓

Vacuum concentration

↓Filtration

Kept in low temperature and deposit the crystal *β*-**CD**

↓

The filtrate was treated with decanol/cyclohexane

↓Filtration

Steam distillation→Retrieved decanol/cyclohexane

↓Drying

α-CD

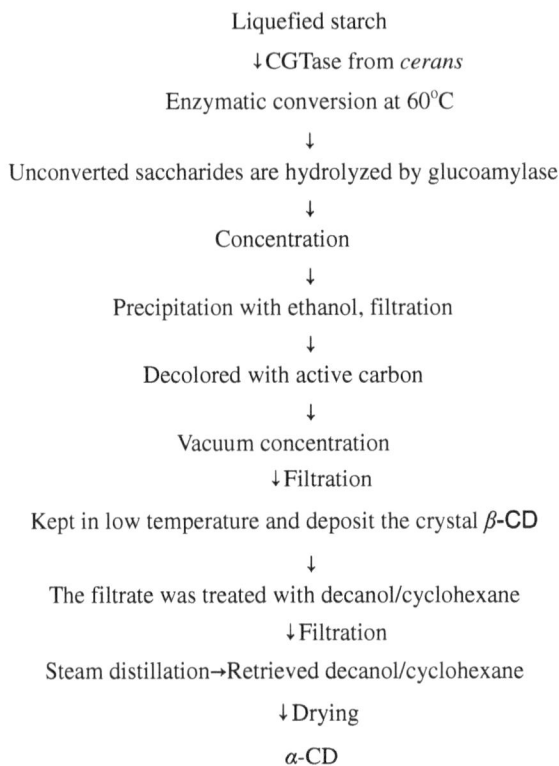

Fig. 3.1.   The preparation of *α*-CD by adding organic solvent.

3.4. The Preparation of *γ*-Cyclodextrin

3.4.1. *Control system*

The process of *γ*-CD preparation by adding organic solvent is shown in Fig. 3.5.

3.4.2. *Non-control system*

The preparation of *γ*-CD without adding organic solvent is shown in Fig. 3.6.

3.5. Preparation of the LR-CDs

LR-CDs are cyclic *α*-1,4-glucans with a DP ranging from nine to hundreds, which were first reported in 1957 [21]. It was not until the 1990's, LR-CDs with a DP of 9–40 were isolated and characterized.

Liquefied starch
    ↓ α-amylase from *B. subtilis*
Enzymatic conversion (4°C)
    ↓

Filtration
    ↓

Ultra-filtration
    ↓

Purification (decoloration and desalination)
    ↓

Concentration
    ↓ Centrifugation
β-CD separation at a low temperature
    ↓

Filtrate
    ↓ α-amylase from *Aspergillus oryzae*
Saccharification
    ↓

Refine and concentration
    ↓ Filtration
Crystal at the low temperature ⟶ Condense the refined filtrate
    ↓   ↓
Crystal α-CD   Starch syrup
    ↓ Drying
    α-CD

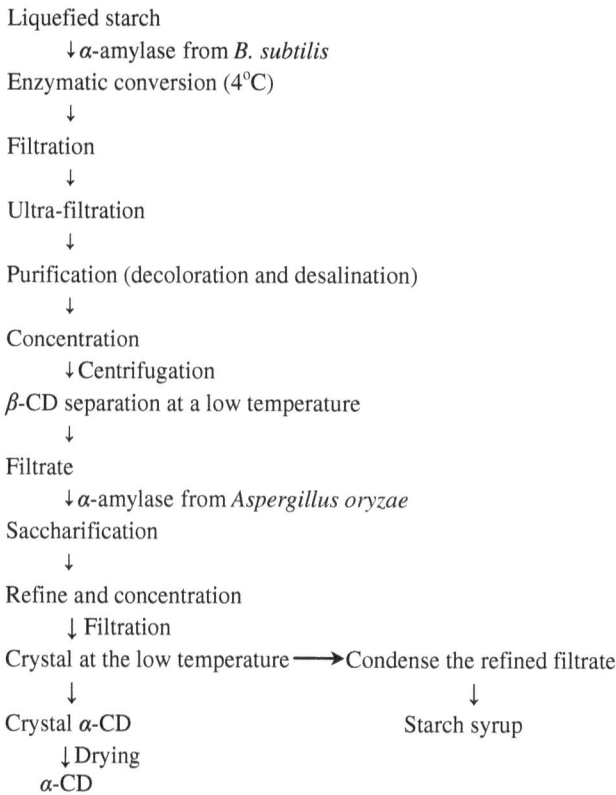

Fig. 3.2. The preparation of α-CD without adding organic solvent.

3.5.1. *Enzyme for LR-CDs preparation*

LR-CDs are formed by various 4-α-glucanotransferases. The DP and yields of LR-CDs depend on the type of enzyme chosen. Takaha *et al.* pointed out that D-enzyme from potato can catalyze transglycosylation to produce LR-CDs with DP ranging from 17 to several hundreds [22]. Terada and Fujii reported that the amylomaltase from *Thermus aquaticus* ATCC33923 can catalyze transglycosylation to form LR-CDs with a yield of 84% and CD_{22} was the smallest LR-CD product [23]. Terada *et al.* found that CGTase enzyme from *Bacillus sp.* A2-5a acted on amylose to form LR-CDs with DP ranging from 9–60 [10]. The CGTase from *B. macerans* can also produce LR-CDs in the initial reaction stages [11]. In addition, Takata *et al.* proved that the branching enzyme (EC 2. 4. 1. 18) from *B. stearothermophilus* can produce LR-CD [24]. Yanase *et al.* found that glycogen debranching enzyme,

Starch

↓α-amylase from *B. subtilis*

Liquefy

↓

Inactive α-amylase by heating

↓CGTase from *B. macerans*

Enzymatic conversion

↓ Add toluent

Stirring

↓

Vacuum filtration

↓

Distill → Retrieve toluent

↓

Vacuum concentration

↓

Purification (decolor and desalt)

↓

β-CD crystallization (Low temperature, 24h)

↓Centrifugation

Wash β-CD using water

↓

Recrystallization

↓

Drying

↓

β-CD

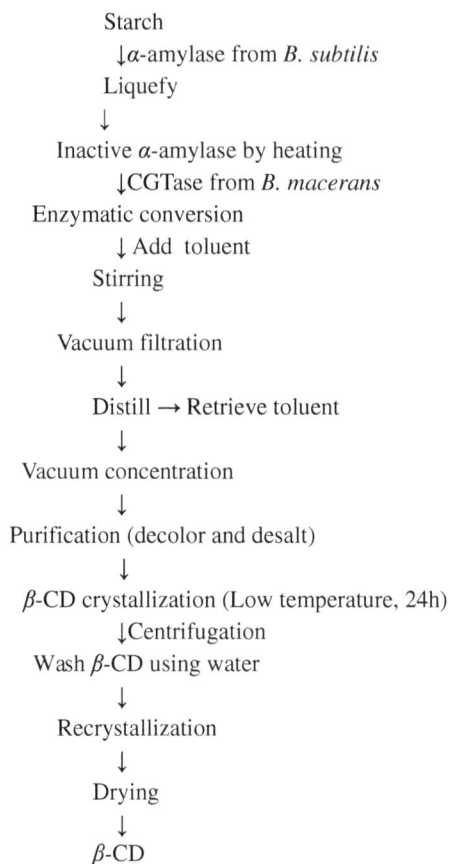

Fig. 3.3.   The preparation of β-CD by adding organic solvent.

obtained from *Saccharomyces cerevisiae* (EC 2. 4. 1. 25/EC 3. 2. 1. 33), can act on amylose and amylopectin to form LR-CDs with DP from 11 to 50 [25].

3.5.2. *Process of LR-CDs preparation*

The process of LR-CD preparation is shown in Fig. 3.7.

3.5.3. *Separation and purification of LR-CDs*

The separation of large ring CD (LR-CD) is very difficult, unlike that of α-, β- and γ-CDs. Based on the existing research results, the separation method is as follows: degrade the unpaired loop dextrin into dextrin or glucose by starch enzymes (such as glucoamylase and debranching enzyme). Dextrin or glucose

Starch liquefy
    ↓CGTase from *Bacillus alkalophilic*
Enzymatic convertion (60°C, 35–45 h)
    ↓Glucoamylase
Mixture of CDs and oligosaccharides
↓
Filtration with membrane (80°C–95°C)
↓
Purification (decoloration and desalination)
↓
Vacuum concentration to 45%–50%
↓
β-CD separation (low temperature, 24 h)
    ↓Centrifugation
β-CD crystalloid (70%–80% retrieve)
↓
Drying
↓
β-CD

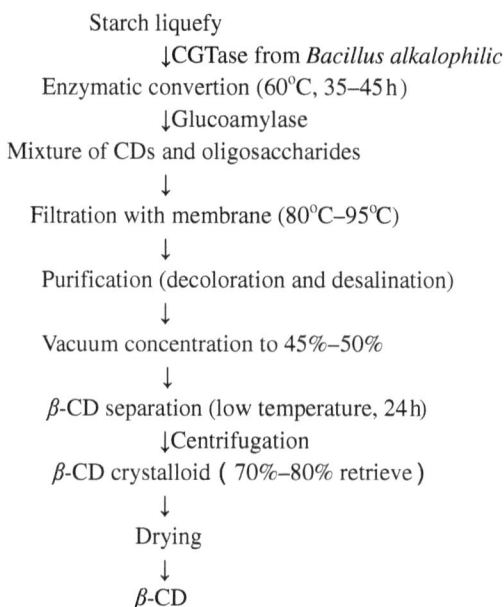

Fig. 3.4. The preparation of β-CD without adding organic solvent.

is then consumed by yeast, and organic solvents are added to precipitate and remove α-, β- and γ-CDs. The remaining CDs are further separated using different chromatographic techniques, especially high-performance ion chromatography (HPAEC). Endo *et al.* reported that LR-CDs with DP of 10–21 were separated by HPAEC [26, 27]. Although scientists have put much effort in LR-CDs separation, there is still a certain distance for LR-CD to be applied in the industry. However, if LR-CD mixtures could be used, rather than LR-CD with a specific DP, the separation costs will be greatly reduced.

3.6. Quantitative and Qualitative Analysis for Cyclodextrin

3.6.1. *Cyclodextrin quantitative analysis*

3.6.1.1. *UV spectrophotometry*

Colored reagents (methyl orange, phenolphthalein and bromocresol green) mixed with CDs would result in an increase or decrease of maximum absorbency, and there is a linearity relation between absorbency and concentration of CDs in the certain range. Thus, UV-spectrophotometry method is widely used for quantitative analysis of CDs.

Starch

$\downarrow \alpha$-amylase from *B. subtilis*

Liquefy

$\downarrow$CGTase from *B. alcalophilus*

Enzymatic convertion (55°C)

$\downarrow$Cyclohexadecane-8-alkene-1-ketone

Stirring for 5–10 h

$\downarrow$

Centrifugation

$\downarrow$

Distill→Retrieve Cyclohexadecane-8-alkene-1-ketone

$\downarrow$

Vacuum concentration

$\downarrow$

Purification (decoloration and desalination)

$\downarrow$

γ-CD separation (low temperature, 24 h)

$\downarrow$Centrifugation

γ-CD hydrated crystalloid

$\downarrow$

Drying

$\downarrow$

γ-CD

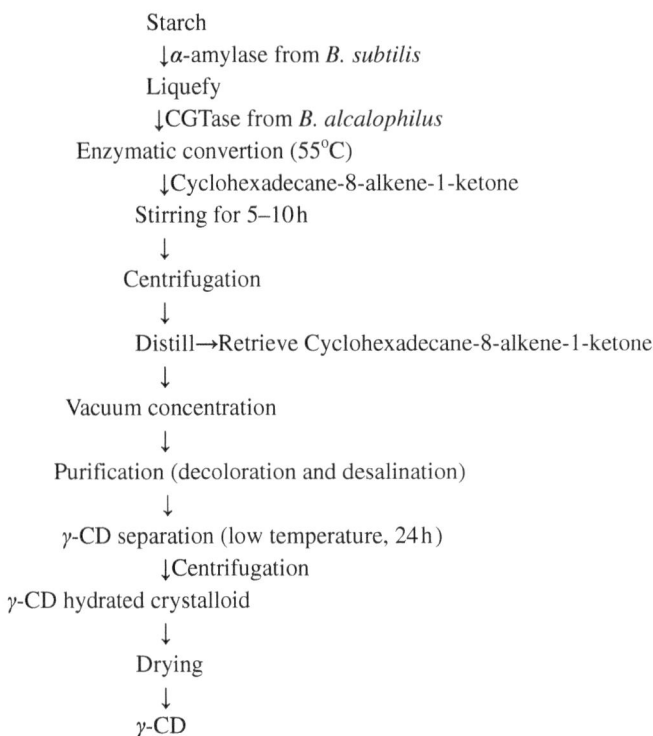

Fig. 3.5. The preparation of γ-CD with organic solvent.

The contents of α-, β- and γ-CDs are measured by UV-spectrophotometry using methyl orange, phenolphthalein and bromocresol green as color reagents, respectively. The detection conditions are as follows:

Quantitative analysis of α-CD: storage solution of methyl orange, 4×10^{-5} mol/L (0.4 mol/L Na$_2$SO$_4$, pH 1.1–1.2); standard α-CD solution, 2×10^{-3} mol/L; series methyl orange solution, 1.0–1.5×10^{-5} mol/L; α-CD solution, 0–1.0×10^{-4} mol/L; $\lambda = 503$ nm.

Quantitative analysis of β-CD: storage solution of phenolphthalein, 3.5×10^{-4} mol/L (0.05 mol/L); boric acid buffer or sodium carbonate buffer, (pH $= 10.5$); standard β-CD solution, 5×10^{-4} mol/L; solution of phenolphthalein, 3×10^{-5} mol/L; series β-CD solution, 0–1.0×10^{-5} mol/L; $\lambda = 553$ nm.

Quantitative analysis of γ-CD: storage solution of bromocresol green, 5×10^{-3} mol/L (20% ethanol); standard γ-CD in citrate buffer (pH $= 4.2$), 5×10^{-4} mol/L; solution of bromocresol green, 1.25×10^{-4} mol/L; series γ-CD solution, 0–2.5×10^{-4} mol/L; $\lambda = 630$ nm.

Starch liquefy
↓CGTase from *B. alkalophilic*
Enzyme converts (60°C, 35-45 h)
↓
Filtration with membrane
↓
Ultrafiltration and concentration
↓
Liquid
↓ Glucoamylase from *Rhizopus sp.*
Concentration
↓
Low temperature →Separate *β*-CD
↓
Centrifugation
↓
Purification using cation exchange resin→Separate glucose
↓
CDs
↓
Purification using gel chromatography→Separate *α*-CD
↓
Concentration and crystallization (low temperature, 24 h)
↓Centrifugation
γ-CD crystalloid
↓
Drying
↓
γ-CD

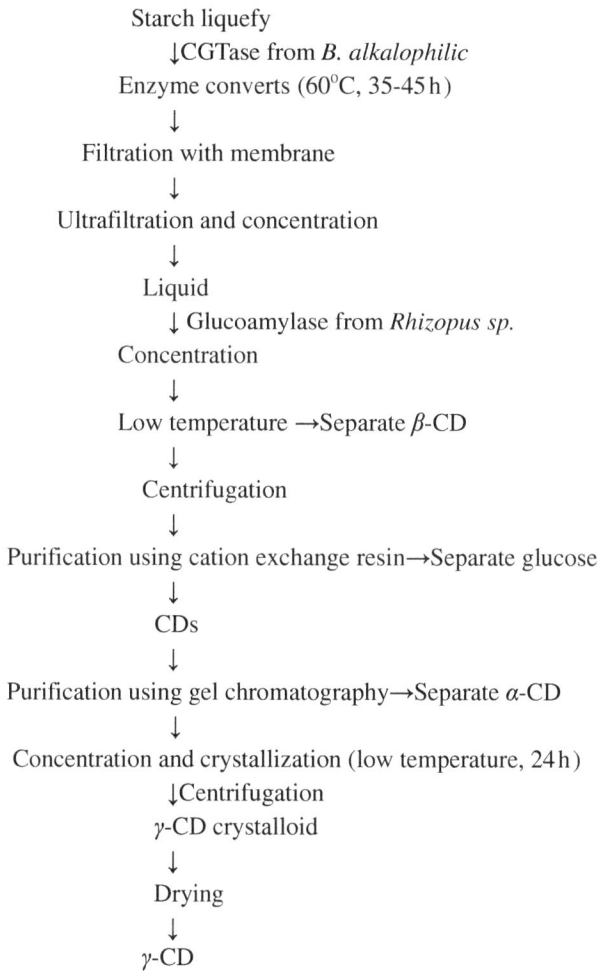

Fig. 3.6. The preparation of *γ*-CD by the non-organic solvent method.

For LR-CD quantification, iodine spectrophotometry could be used because of the LR-CD-iodine complexes formation. For LR-CD mixture with DP range of 22–40, the detection maximum wavelength could be set as 496 nm, using 0.02 M I/KI solution as reagent.

3.6.1.2. *High performance liquid chromatography*

High-performance liquid chromatography (HPLC) with differential refractometer (RI) detector could be used for CD quantification. When using the NH_2-column,

Starch

↓

CGTase

↓

Inactive

↓

Filtrate

↓

Glucoamylase ⟶ Unconverted starch

↓

Inactive

↓

Concentration

↓

Acetone precipitate ⟶ Desalt and discard glucose

↓

Wash precipitate using acetone

↓

Drying

↓

LR-CD

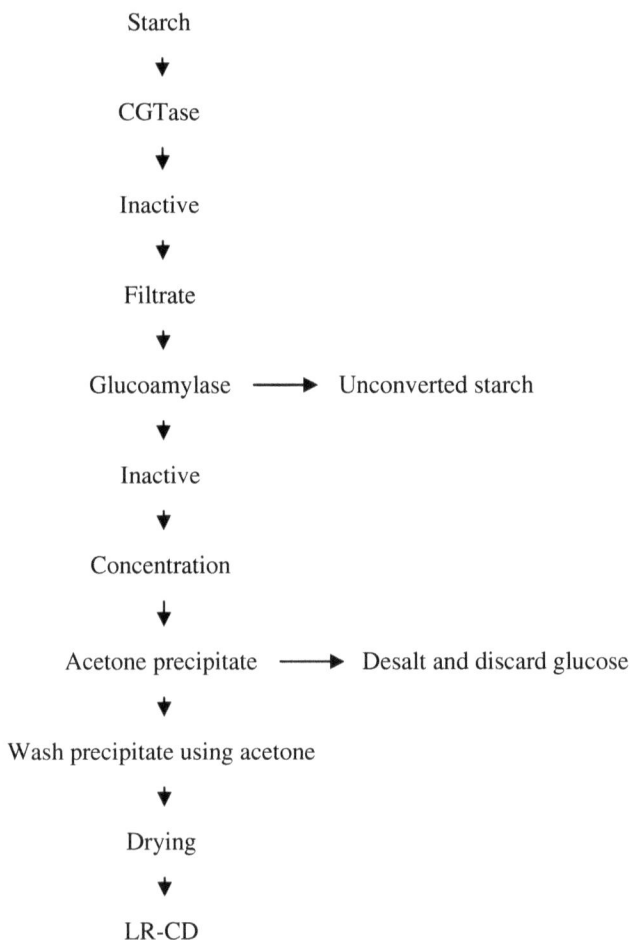

Fig. 3.7.   The preparation of LR-CD.

the eluant was acetonitrile/water with the concentration ranging from 70/30 to 60/40 (V/V). When using C-18 column, the eluant was metanol/water with concentration ranging from 4/96 to 10/96 (V/V). Thus, CDs could be quantified by external standard method.

3.6.2. *Cyclodextrin qualitative analysis*

3.6.2.1. *Paper chromatography and thin layer chromatography*

Paper chromatography and thin layer chromatography (TLC) are the rapid and simple methods for qualitative analysis of α-, β- and γ-CDs and these two methods

Table 3.2. The general conditions of CDs measured by paper and thin layer chromatography.

Chromatography	Developer	Coloration
Paper chromatography	Butanol/pyridine/H_2O (6:4:3) Butanol/N,N-imethylformamide/H_2O (2:1:1) 65% propanol (60°–70°C)	Iodine dissolved in 80% acetone
Silica chromatography	Butanol/acetic acid/pyridine/H_2O/ N,N-Dimethylformamide (6:3:1:2:4) Butanol/acetic acid/H_2O (6:3:1)	50% of ethanolic sulphuric solution
Microcrystalline cellulose chromatography	Butanol/ethanol/H_2O (4:3:5)	

are commonly used in the industry for CD identification. The general conditions of CDs measured by these two methods are shown in Table 3.2.

TLC has been used for qualitative analysis of LR-CD. The panel is made by special sorbents: Kieselgel 60 and modified NH_2-Kieselgel 60. The developer for Kieselgel 60 is dioxane/aqueous ammonium solution (25%, 1:1), and the plate is developed twice; while the developer for NH_2-Kieselgel 60 is acetonitrile/water (55:45). About 50% of ethanolic sulphuric solution is used as coloration.

3.6.2.2. *Gas chromatography*

CDs cannot be detected directly by gas chromatography (GC), except derived by dimethyl-silicon-ether beforehand. The analysis conditions are as follows: Column: 3% SXR covered with Chromosorb WAW DMCS (80/100 mesh); temperature program: 325°C–405°C, 20°C/min; Carrier gas (He) velocity: 45–50 mL/min. It is crucial for accurate analysis to control the derivative process.

3.6.2.3. *Capillary electrophoresis*

Since CDs are colorless, without any chromophore group and only electrified at a very high pH. A direct separation and detection of CDs on capillary electrophoresis (CE) using detectors of UV and laser induced fluorescence (LIF) is impossible. With special inclusion capabilities, CDs could form inclusion complexes with a large range of aromatic ions, which would provide UV absorption or emit fluorescence, and is able to facilitate the analysis of CDs using CE. CDs with DP of 6–13 had been analyzed and characterized using CE [31]. The reported

Table 3.3.   Inclusion complex formations of CDs as single or mixtures detected by CE.

CD	Compound	References
CD6–CD8	Benzoate, 2-anilinonaphthalene-6-sulfonic acid, salicylic acid, benzylamine	28
CD9	Dansyl-methione, dansyl-phenylalanine, dansyl-alanine, dansyl-leucine, dansyl-norvaline, dansyl-tryptophan, dansyl-glutamic acid, dansyl-aspartic acid, dansyl-threonine, 1,1′-binaphthyl-2,2′-diylhydrogenphosphate, carvedilol, erythro-mefloquine, clidinum bromide, FMOC-phenylalanine, FMOC-tryptophane, FMOC-alanine, narigin, hespesperetin, neohesperidin, neoeriocitrin	29
CD9–CD13	Benzoat, 2-methyl benzoate, 3-methyl benzoat, 3-methyl benzoate, 4-methyl benzoate, 2,4-dimethyl benzoate, 2,5-dimethyl benzoate, 3,5-dimethyl benzoate, 3,5-dimethoxy benzoate, salicylate, 3-phenyl propionate, 4-tert-butyl benzoate, ibuprofen anion, 1-adamantane carboxylate	28
CD14–CD17	Salicylate, 4-tert-butyl benzoate, ibuprofen anion	30
CD22–CD40	Iodine	31

complex formations of CDs as single components and as mixture detected by CE are summarized and listed in Table 3.3.

3.6.2.4. *High performance anion-exchange chromatography with pulsed amperometric detection*

In high performance anion-exchange chromatography with pulsed amperometric detection (HPAEC-PAD) system, anion exchange column is used for CD separation. CarboPac PA-10 and 100 are all selected for CD qualitative analysis. The main eluent for the column of CarboPac PA-100 is sodium hydroxide, the gradient eluent is sodium acetate.

3.6.2.5. *Mass spectrometry and nuclear magnetic resonance*

Mass spectrometry (MS) and nuclear magnetic resonance (NMR) are most widely used technologies for CDs identification. Fast atom bombardment mass spectrometry (FAB-MS), matric-assisted laser desorption ionization time-of-flight mass spectrometry (MALDI TOF MS) are used for CDs molecular weight detection, from which to identify the DP range of CD. NMR is used for CDs structure identification.

References

1. Nakagawa, T, K Ueno, M Kashiwa and J Watanabe (1994). The stereoselective synthesis of cyclomaltopentaose. A novel cyclodextrin homologue with D.P. five. *Tetrahedron Letters*, 35, 921–1924.

2. Wakao, M, K Fukase and S Kusumoto (2002). Chemical synthesis of cyclodextrins by using intramolecular glycosylation. *Journal of Organic Chemistry*, 67, 8182–8190.

3. Kim, TJ, YD Lee and HS Kim (1993). Enzymatic production of cyclodextrins from milled corn starch in an ultrafiltration membrane bioreactor. *Biotechnology and Bioengineering*, 41, 88–94.

4. Nakano, H and S Kitahata (2005). Application of cyclodextrin glucanotransferase to the synthesis of useful oligosaccharides and glycosides. In *Handbook of Industrial Biocatalysis*, T Hou Ching (ed.), pp. 22-1–22-17. CRC Press.

5. Cao, XZ, ZY Jin, F Chen and X Wang (2004). Purification and properties of cyclodextrin glucanotransferase from an Alkalophilic bacillus sp.7-12. *Journal of Food Biochemistry*, 28, 463–475.

6. Zhang, XP, P Zhang and CL Li (1994). Properties of cyclomaltodetrin-glucanotransferase and production of cyclodextrin. *Acta Scientiarum Naturalium Universituns Nakaienses*, 12, 63–67. (in Chinese)

7. Bender, H (1997). Cyclodextrin–glucanotransferase von *Klebsiella pneumoniae*, 1. Synthese, Reinigung und Eigenschaften des Enzyms von K. pneumoniae M5a1. *Archieve of Microbiology*, 111, 271–282.

8. Kobayashi, S, K Kainuma and S Suzuki (1978). Purification and some properties of *Bacillus macerans* cycloamylose (cyclodextrin) glucanotransferase. *Carbohydrate Research*, 61, 229–238.

9. Gawande, BN and AY Patkar (1999). Application of factorial designs for optimization of cyclodextrin glycosyltransferase production from *Klebsiella pneumoniae* AS-22. *Biotechnology and Bioengineering*, 64, 168–173.

10. Terada, Y, H Sanbe and T Takaha (2001). Comparative study of the cyclization reactions of three bacterial cyclomaltodextrin glucanotransferase. *Applied and Environmental Microbiology*, 67, 1453–1460.

11. Larsen, KL, HJS Christensen and F Mathiesen (1998). Production of cyclomaltononaose (δ-cyclodextrin) by cyclodextrin glycosyltransferaases from Bacillus sp. and bacterial isolates. *Applied and Environmental Microbiology*, 50, 314–317.

12. Koizumi, K, H Sanbe and Y Kubota (1999). Isolation and characterization of cyclic [alpha]-(1–>4)-glucans having degrees of polymerization 9-31 and their quantitative analysis by high-performance anion-exchange chromatography with pulsed amperometric detection. *Journal of Chromatography A*, 852, 407–416.

13. Kato, T and K Horikosgik (1986). Cloning and expression of the *Bacillus subtilis* No. 313 γ-cyclodextrin forming CGTase gene in *Escherichia coli*. *Agricultural and Biological Chemistry*, 50, 2161–2162.

14. Eglbrecht, A, G Harrer, M Lebert, G Schmid and D Duchene (1990). *Minutes of the Fifth International Symposium on Cyclodextrins*, pp. 25–31. Paris: Editions de Santé.

15. Fujita, Y, H Tsubouchi, Y Inagi, K Tomita, A Ozaki and K Nakanishi (1990). Purification and properties of cyclodextrin glycosyltransferase from *Bacillus sp.* AL-6 *Journal of Fermentation Bioengineering*, 70, 150–154.

16. Pongsawadi, P and M Yagisawa (1987). Screening and identification of a cyclomaltox-trinv glucanotransferase-producing bacteria. *Journal of Fermentation Technology*, 65, 463–467.

17. Abelyan, VA, AM Balayan, LS Manukyan, KB Afyan, VS Meliksetyan, NA Andreasyan and AA Markosyan (2002). Characteristics of cyclodextrin production using cyclodextrin glucanotransferases from various groups of microorganisms. *Applied Biochemistry and Microbiology*, 38, 527–535.

18. Qi, Q, X She, T Endo and W Zimmermann (2004). Effect of the reaction temperature on the transglycosylation reactions catalyzed by the cyclodextrin glucanotransferase from *Bacillus macerans* for the synthesis of large-ring cyclodextrins. *Tetrahedron*, 60, 799–806.

19. Endo, T, N Ogawa, H Nagase, H Sambe, T Takaha, Y Terada, W Zimmermann and U Haruhisa (2007). Production of large-ring cyclodextrins composed of 9-21α-D-glucopyranose units by cyclodextrin glucanotransferase-effects of incubation temperature and molecular weight of amylose. *Heterocycles*, 74, 991–994.

20. Chen, MN (1995). *Starch and Starch Sugar*, 2, 50–52. (in Chinese)

21. French, D, AO Pulley and JA Effenberger (1965). Studies on the Schardinger dextrins. XII. The molecular size and structure of the δ-, ε-, ζ-, and η-dextrins. *Archives of Biochemistry Biophysics*, 111, 153–160.

22. Takaha, T, M Yanase and S Okada (1996). Glucan having cyclic structure and its production. Japanese Patent 08-311103.

23. Terada, Y and K Fujii (1999). *Thermus aquaticus* ATCC 33923 amylomaltase gene cloning and expression and enzyme characterization: Production of cycloamylose. *Applied and Environmental Microbiology*, 65, 910–915.

24. Takata, H, T Takaha and S Okada (1996). Cyclization reaction catalyzed by branching enzyme. *Journal of Bacteriology*, 178, 1600–1606.

25. Yachibana, Y, T Takaha and S Fujiwara (2000). Acceptor specificity of 4-α-glucanotransferase from *Pyrococcus kodakaraensis* KOD1, and synthesis of cycloamylose. *Journal of Bioscience and Bioengineering*, 90, 406–409.

26. Endo, T, H Nagase and H Ueda (1997). Isolation, purification and char-acterization of cyclomaltotetradecaose (ι-cyclodextrin), cyclomaltopentadecaose (κ-cyclodextrin), cyclomaltohexadecaose (λ-cyclodextrin) and cyclomaltoheptadecaose (μ-cyclodextrin). *Chemical and Pharmaceutical Bulletin*, 45, 1856–1859.

27. Endo, T, H Nagase and H Ueda (1998). Isolation, purification and characterization of cyclomaltooctadecaose (ν-cyclodextrin), cyclomaltononadecaose (ξ-cyclodextrin), cylomaltoeicosaose (o-cyclodextrin) and cyclomaltoheneicosaose (π-cyclodextrin). *Chemical and Pharmaceutical Bulletin*, 6, 1840–1843.

28. Larsen, KL and W Zimmermann (1999). Analysis and characterisation of cyclodetxrins and thier inclusion complexes by affinity capillary electrophoresis. *Journal of Chromatography A*, 836, 3–14.

29. Wistuba, D, A Bogdanshi, KL Larsen and V Schuring (2006). δ-cyclodextrin as novel chiral probe for enantiomeric separation by electromigration methods. *Electrophoresis*, 27, 4359–4363.

30. Mogensen, B, T Endo, H Ueda, W Zimmermann and KL Larsen (2000). Analysis of complex formation between large-ring cyclodextrins and aromatic anions. *In cyclodextrin: From Basic Research to Market. 10th International Cyclodextrin Symposium,* Ann Arbor, MI, USA: CD-ROM edition.
31. Wang, JP, AQ Jiao, YQ Tian, XM Xu and ZY Jin (2011). Isolation of cycloamylose by iodine affinity capillary electrophoresis. *Journal of Chromatography A*, 1218, 863–868.

4

PREPARATION OF BRANCHED-CYCLODEXTRINS

Xing Zhou, Yao-Qi Tian and Zheng-Yu Jin

The State Key Laboratory of Food Science and Technology
School of Food Science and Technology, Jiangnan University,
Wuxi 214122, China

The so-called first-generation cyclodextrin (CD), i.e. the native α-CD, β-CD and γ-CD, each has its limitations in application. β-CD exhibits poor solubility; α-CD owns a small cavity which restricts the size of the guest molecules; γ-CD has a larger cavity but expensive. Thus, to expand the application area, native CDs are usually modified to improve their physiochemical properties [1]. Until now, various branched-CDs, such as methyl-CD, hydroxypropyl-CD, glucosyl-CD (G1-CD) and maltosyl-CD (Mal-CD) have already been commercialized.

Generally, there are two methods to produce branched-CDs: chemical method and enzymatic method. Scientists first discovered the enzymatic method and used it in applications. However, because of the low conversion rate and high cost of the enzymatic method, the chemical method was developed gradually. The chemical method has its own drawbacks: the isolation and purification processes are very complex. Therefore, the enzymatic method is still the mainstream to produce branched-CDs. During 1980s' and 1990s', the main products of the enzymatic modified CDs were the homobranched CDs which are called second-generation CDs. From the early 1990s', researchers began to investigate the third generation CDs, i.e. the heterobranched CDs modified by various glycosyl groups.

The branched-CDs exhibit much better properties as compared with native CDs.

- **Solubilizing and stabilizing ability.** Branched-CDs have better water solubility and can form soluble inclusion complex. The commercial branched-CDs-essence inclusion complex is stable at 90°C for 7 h without damages; whereas only 45% of essence in the CDs-essence complex remains at the same condition. Branched-CDs-fatty acid complex, which improves the solubility of fatty acid in water, can be applied in serum-free culture for animal cell and cancer cell.
- **Low hemolytic activity.** Investigations have demonstrated that the hemolytic activity of 6-O-α-(4-O-α-D-glucuronyl)-D-glucosyl-β-CD (GUG-β-CD) on the rabbit red cell is not only lower than that of native β-CD but also lower than that of maltosyl-β-CD (Mal-β-CD). The inclusion capacity of GUG-β-CD for acidic drugs and neutral drugs is equal to or a bit lower than that of native β-CD and Mal-β-CD, but the electrostatic interaction between negative ion of GUG-β-CD carboxylic group and positive ion of essential drugs makes GUG-β-CD show higher affinity to essential drugs. Thus, for the essential drugs, GUG-β-CD is a safe solvent [2].
- **Target carrier.** Galactosyl-CDs (Gal-CDs) are possible to act as the hepatocellular drug target carrier. 10 mM Gal-CDs or galactose can inhibit the combination of a lactose-carrying styrene homopolymer (PVLA) and hepatocyte but 20 mM glucose or G1-CDs cannot.

Because of the improved properties, branched-CDs have much broader applications as in food, medicine and cosmetics [3]. As the drug carriers, branched-CDs show little hemolytic activity and muscle irritation. They can be used to increase solubility, improve bioavailability and stability of chiral drugs and reduce side effects of drugs. In the field of food and spices, branched-CDs can be more effective in preventing the loss of flavor due to light, heat, oxidation and volatilization, improving the stability of the nutrients and masking or correcting the bitter/strange taste of food ingredients. In the cosmetics field, branched-CDs can reduce the stimulation of the organic ingredients to the skin mucosal tissues, and enhance the stability and prevent oxidation and evaporation of effective ingredients. The application fields of branched-CDs have been gradually extended and are used in pesticides, new materials, environmental protection, etc [1].

In this chapter, preparation, isolation and identification of the branched-CDs, i.e. Mal-CDs and Gal-CDs, will be discussed in detail, and the preparation of G1-CD, Man-G1-CD, Gal-G1-CD and Gal-Mal-CD will be briefly mentioned.

4.1. Basic Theories

4.1.1. *Synthesis mechanisms*

Branched-CDs can be synthesized from native CDs or even monosaccharide and oligosaccharide through the action of different enzymes. The type of branch also depends on the enzyme. Mal-CD is synthesized by the condensation reaction of high concentrated CDs and maltose through the action of pullulanase or isoamylase. Gal-CDs can be produced from melibiose and CDs. The reaction is catalized by α-galactosidase. G1-CDs can be produced from melibiose and CDs, which is catalyzed by cyclodextrin glucanotransferase (CGTase). Mannosyl-CDs can be produced from α-mannopyranoside and CDs, catalyzed by α-mannosidase.

4.1.2. *Using CD and maltodextrin/starch as raw material*

G1-CD was discovered in 1965 and proved at the end of 1966. However, the yield of G1-CD was very low, i.e. 2% (amylopectin as substrate), and difficult to separate. Later on, based on the study of CGTase from *Bacillus macerans*, it was established that waxy corn starch could be used as the starting material to produce G1-CD. This method was improved in 1984 when the G1-CD crystalline was obtained and the structure and properties of G1-CD were characterized.

4.1.3. *Using CD and α-maltosyl fluoride (α-G2F) as raw material*

Alpha-maltosyl fluoride (α-G2F) is an active compound [4]. When 40 mM α-G2F and 90 mM α-CD are incubated in 2.8 U/mL pullulanase at 60°C for 1 h, the yield of Mal-α-CD is 12 mM. The relative conversation rate is 32%. Whereas, the yield of Mal-α-CD is only 0.5 mM when α-G2F is replaced by α-maltose under the same condition.

4.1.4. *The preparation of galactosidase and enzymatic preparation of Gal-CD*

Alpha-galactosidase, originated from green coffee beans, can directly transfer galactose residues to α-CD. It is easy to extract and purify coffee bean α-galactosidase, however, the low yield and activity restricted its application and development. Recently, it was found that α-galactosidase extracted from germinated coffee beans exhibited higher activity than that from coffee beans. Germinated coffee beans are readily available and they can become a new source for α-galactosidase. After adding coffee bean α-galactosidase to the buffer containing melibiose and CDs (α-, β-, γ-CDs) and incubating at 40°C for 48 h, the reaction system is composed of mixtures of Gal-CDs (Gal-α-CD, Gal-β-CD and Gal-γ-CD), melibiose, and CDs and hydrolyzate of melibiose. The reaction mixture can be separated to obtain Gal-α-CD, Gal-β-CD and Gal-γ-CD using preparative HPLC (Asahipack NH2P-50). These Gal-CDs can be hydrolyzed to be equal moles of galactose and corresponding CDs. Fast atom bombardment-mass spectrometry (FAB-MS) and ^{13}C-NMR analysis suggests that these Gal-CDs are 6-O-Galactofuranose-α-CD, 6-O-Galactofuranose-β-CD and 6-O-Galactofuranose-γ-CD.

4.2. Preparation of Mal-CDs

Maltosyl (α-1→6) β-CDs (Mal-β-CD) is a kind of β-CD derivative. Its solubility in water is 151 g/100 mL, much higher than that of native β-CD (1.85 g/100 mL). Mal-β-CD and other β-CD derivatives have recently become a research focus in the world. Shaojie Wang and Zhengyu Jin [5] and Bo Cui and Zhengyu Jin [6, 7] studied the methods for preparation and analysis of Mal-β-CD. Bo Cui and Zhengyu Jin investigated the structure of Mal-β-CD [8].

4.2.1. *Reaction conditions*

Usually, Mal-β-CD is synthesized by the reverse reaction of pullulanase (EC 3.2.1.41) and isoamylase (EC 3.2.1.68) using maltose and β-CD as substrate [9], or by the transfer reaction of pullulanase which transfers α-maltosyl fluoride to β-CD [4, 10]. Considering the yield and safety, the former method is better. Shaojie Wang *et al.* chose *Bacillus licheniformis* pullulanase which is resistant to acid and heat to reversely synthesize Mal-β-CD [11]. Pullulanase is commonly used to hydrolyze 1, 6-glycosidic bond of starch side chains. But when the enzyme concentration and substrate concentration are high, the reaction can be reversed. The mechanism is shown in Fig. 4.1.

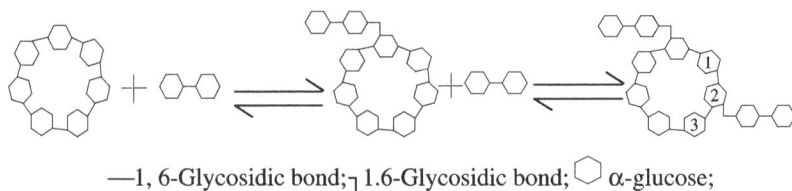

—1, 6-Glycosidic bond;⌐ 1.6-Glycosidic bond; ◡ α-glucose;

Fig. 4.1.   Schematic diagram of reverse synthesis Mal-β-CD by pullulanase [12].

4.2.1.1. *Pullulanase activity*

Pullulanase exhibits highest activity at pH 5 and is stable at pH 5–6. It shows relatively higher activity when the temperature is around 50°C–60°C, and is stable at 30°C–60°C, but when the temperature is higher than 60°C, it will be rapidly inactivated [6].

4.2.1.2. *Optimization of the production of Mal-β-CD by pullulanase*

4.2.1.2.1. pH value

The optimal pH of pullulanase is around 5.0 for the hydrolysis reaction, whereas the optimal pH of pullulanase for the reverse synthesis reaction is around 4.0–4.5, as shown in Fig. 4.2.

4.2.1.2.2. Temperature

Temperature can affect the catalytic reaction activity. At a certain temperature range, the increasing of temperature increased the enzyme activity. However, further increasing of the temperature might destroy the conformation of the enzyme, and may result in a drop of the conversion rate. On the other hand, the solubility of β-CD and maltose can be improved and the viscosity of the reaction system can be reduced by increasing the temperature. The optimal temperature of pullulanase is around 60°C–80°C [6].

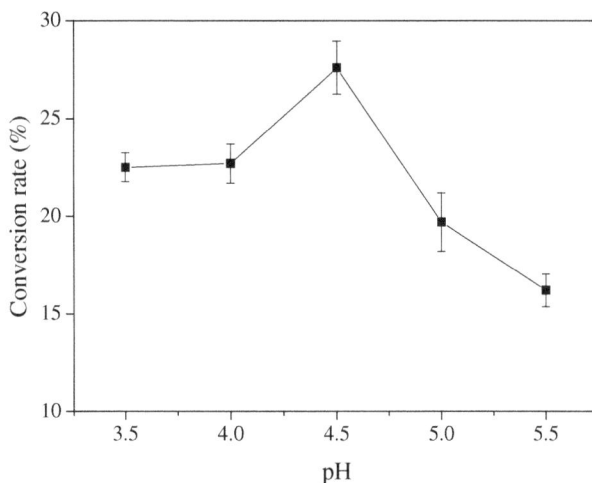

Fig. 4.2. Effect of pH on the conversion rate of Mal-β-CD. Reaction conditions: mole ratio between maltose and β-CD: 16:1, substrate concentration of 70%, 100 U pullulanase per gram of β-CD, temperature of 60°C and reaction time of 60 h [6].

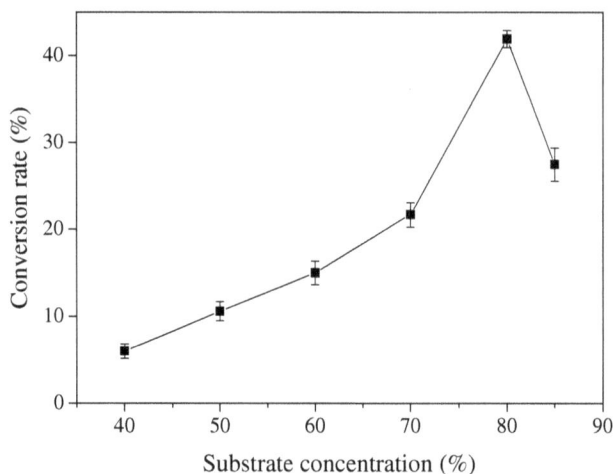

Fig. 4.3.   Effect of substrate concentration on the conversion rate of Mal-β-CD. Reaction conditions: the mole ratio of maltose and β-CD: 16:1, 100 U pullulanase per gram of β-CD, the temperature of 60°C, reaction time of 60 h and pH = 4.0 [6].

4.2.1.2.3. Substrate concentration

At low substrate concentration, the hydrolysis reaction of pullulanase plays a dominant role.

Only high substrate concentration is favorable for the reverse synthesis reaction of pullulanase (Fig. 4.3). But when the substrate concentration is higher than 80%, the conversion rate of Mal-β-CD starts to decrease. It might be because the viscosity of the reaction system becomes too high, which is harmful for the enzymatic reaction, when increasing the substrate concentration to above 80% [6].

4.2.1.2.4. The ration between β-CD and maltose

The ratio between β-CD and maltose also plays an important role. β-CD has a poor solubility but maltose is well soluble. Thus, the conversion rate is increased with an increase with an increase in the mole ratio between maltose and β-CD. The conversion rate reaches very high levels when the ratio between maltose and β-CD is higher than 19 (Fig. 4.4).

4.2.1.2.5. The amount of pullulanase

The amount of pullulanase is the most influential factor of the conversion rate of β-CD [6]. As shown in Fig. 4.5, when the amount of pullulanase is increased from

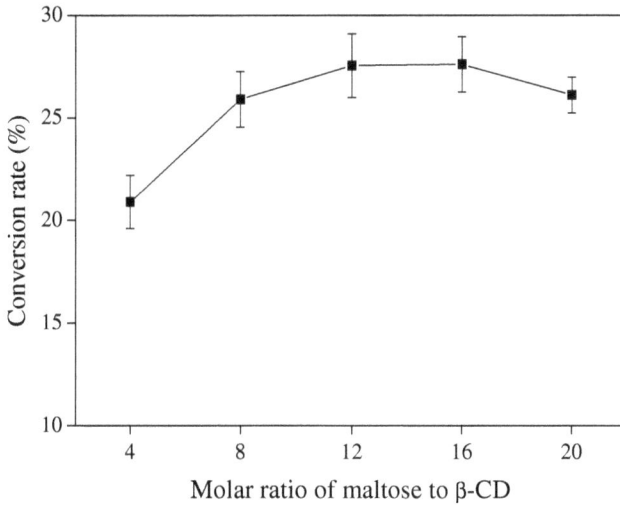

Fig. 4.4. Effect of the mole ratio between maltose and β-CD on the conversion rate of Mal-β-CD. Reaction conditions: substrate concentration of 70%, 100 U pullulanase per gram of β-CD, the temperature of 60°C, reaction time of 60 h and pH $= 4.0$ [6].

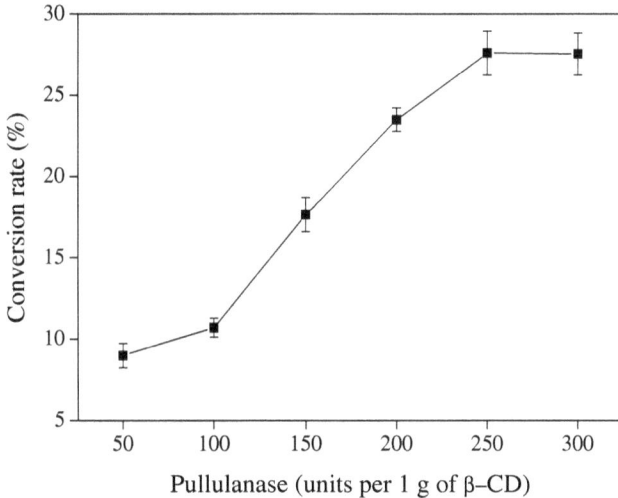

Fig. 4.5. Effect of the amount of pullulanase on the conversion rate of Mal-β-CD. Reaction conditions: the mole ratio of maltose and β-CD: 16:1, substrate concentration of 70%, the temperature of 60°C, reaction time of 60 h and pH $= 4.0$ [6].

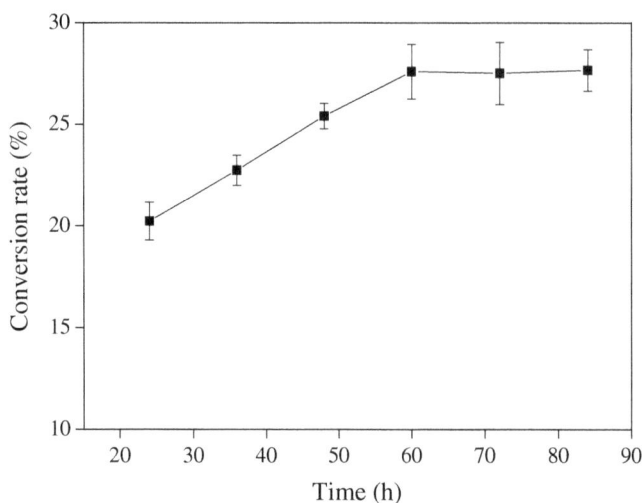

Fig. 4.6. Effect of reaction time on the conversion rate of Mal-β-CD. Reaction conditions: the mole ratio of maltose and β-CD: 16:1, substrate concentration of 70%, 100 U pullulanase per gram of β-CD, the temperature of 60°C, and pH = 4.0 [6].

60 U/g to 100 U/g β-CD, the conversion rate increases quickly. The conversion rate shows slow increase when the enzyme amount is higher than 200 U/g β-CD.

4.2.1.2.6. Reaction time

The production of Mal-β-CD by the reverse synthesis reaction needs a long time (Fig. 4.6). But considering the cost, reducing enzyme activity and the darkening of the product due to long time reaction, the reaction time should not be too long.

According to Bo Cui, the overall optimal conditions of the production of Mal-β-CD by the reverse synthesis reaction of pullulanase, obtained by orthogonal experimental design are: temperature of 70°C, pH of 4.2, substrate ratio of 20 and enzyme amount 300 U/g β-CD [6].

4.2.2. *Isolation and purification of Mal-β-CD*

The crude Mal-β-CD product contains: maltose, β-CD, Mal-β-CD and a little branched-tetrose. Thus, membrane filtration technology, i.e. nanofiltration (NF), and chromatographic method can be utilized to isolate Mal-β-CD from crude Mal-β-CD.

Membrane filtration technology has high selectivity, requires simple equipment and low energy cost, can be operated at room temperature, and during the whole process there are no phase and chemical changes. The combination of appropriate membrane, operating parameters and the separation mode can achieve the concentration, separation and purification of the material.

4.2.2.1. *NF*

The nominal pore size of the NF membrane is typically about 1 nm. NF can retain molecules which have size of 100–1000 nm which ranges between ultrafiltration and reverse osmosis. The relative molecular weight of Mal-β-CD is 342. The NF membrane (NE-1812) with molecular weight cut off (MWCO) of 200–400 can be used to initially concentrate the dilute solution of crude Mal-β-CD. There are two important indicators: membrane flux and rejection coefficient of maltose in the crude Mal-β-CD.

The mechanism of NF is similar as that of reverse osmosis separation, following the same formula:

$$JW = A \times (\Delta P - \Delta X),$$

$$JS = B \times \Delta C.$$

JW and JS stand for the solvent and solute membrane flux, respectively. A and B are the parameters related with the nature of the membrane material. ΔP, ΔX and ΔC, stand for the pressure difference, the osmotic pressure difference and the solute concentration difference between inside and outside of the membrane, respectively. The basic principle is to use the selective permeability of polymer membrane and the driving force of the concentration gradient, pressure gradient, the osmotic pressure gradient to transfer mass between the membrane inter-phase to achieve separation and purification of different components. Inorganic salts can pass through NF membrane. The osmotic pressure of NF membrane is lower than the RO membrane.

As can be seen from the equation, when the flux is constant, NF requires lower external pressure than reverse osmosis do, while the same pressure, the flux of NF is greater than reverse osmosis. This is the reason why the operating pressure NF possesses the quality of low operating pressure and high flux.

The diagram of NF device is shown in Fig. 4.7. During NF, a major problem is that the membrane flux drops with the extension of operation time [13].

Table 4.1 lists the water permeability and the retention rate of single-component sugars of NE-1812 membrane under different operating pressures. Under certain operating conditions, NE-1812 membrane is very stable and its pure water permeability is almost unchanged, average of 15.5×10^{-6} ms$^{-1} \bullet$ MPa^{-1}, with the operating pressure. When the operating pressure is 0.5 MPa, the retention rate of glucose by NE-1812 membrane is 80%, that of maltose is 93% maltose and that of raffinose is 99%, indicating that MWCO of NE-1812 membrane is about 300.

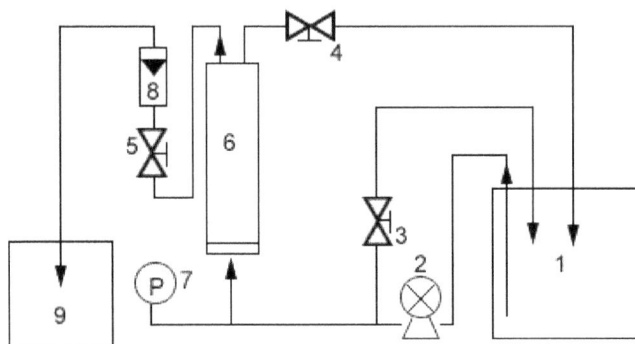

Fig. 4.7. NF devices. 1. Raw material storage tank; 2. Infusion pump; 3. Circulating valve; 4. Concentration valve; 5. Flowmeter valve; 6. Membrane components; 7. Gauge; 8. Flowmeter; 9. Penetrant storage tank [12].

Table 4.1. Pure water permeability of NE-1812 and the retention rate of monosaccharide and sodium chloride by NE-1812 membrane under different operating pressures [12].

Pressure (MPa)	Pure water permeability $Lp(10^{-6} m.s^{-1} \cdot MPa^{-1})$	Robs (glucose)	Robs (maltose)
0.2	15.48	0.78	0.92
0.3	15.43	0.80	0.92
0.4	15.52	0.81	0.93
0.5	15.53	0.82	0.95
0.6	15.52	0.82	0.94

Temperature: $30°C$, the concentration of each monosaccharide: $0.2 g/L$.

By increasing the pressure, membrane pores become smaller because of the compaction effect, resulting in the increase of retention rate. Sarney *et al.* [13] and Aydogan *et al.* [14] also reported that the effect of pressure on the retention rate of sugar component was not significant. Bo Cui found that the pressure can be maintained at about 0.5 MPa, as long as the pressure is in the range of membrane tolerance [12].

The higher the temperature, the greater is the average membrane fluxes. The maximum operating temperature of NE-1812 membrane is $50°C$. But long time operation at a high temperature $(50°C)$ is harmful to the membrane. Thus, the operating temperature is better to be controlled at about $40°C$.

According to Bo Cui *et al.*, when the maltose concentration is around $2–10 g/L$, there is little change in membrane flux and maltose rejection rate; whereas when the maltose concentration is increased to 10 g/L, the permeation flux decreases significantly [12].

The overall optimized operation conditions are as follows: pressure of 0.5 MPa, temperature of 40°C and maltose concentration of 10 g/L.

4.2.2.2. *Active carbon separation*

Active carbon's, absorption capacity as a solid absorbent was affirmed for the first time in 1773, patented in 1990 and its products have since been widely used. Active carbon has many advantages: extensive sources, simply producible, large surface area and high absorption speed. As a nonpolar solid adsorbent, it can easily adsorb organic matter and molecules with small polarity.

Bo Cui separated Mal-β-CD by gradient elution, i.e. using 10%, 20%, 40% and 60% ethanol, from active carbon column [12]. 10% ethanol eluate contains almost all malt, 20% ethanol eluate contains about 56% maltose, 40% ethanol eluate contains only 1.5% maltose, and 60% ethanol eluate contains about 18% of maltose. With the increase of ethanol concentration, the elution capacity increases. 40% ethanol eluate contains only 1.5% maltose, i.e. the content of β-CD and Mal-β-CD are as high as 98.5%, but when ethanol concentration reaches 60%, the amount of maltose in the elution increases.

4.2.2.3. *Separation using Sephadex G-15*

Gel chromatography was developed during the late 1950s and early 1960s. It is a quick and easy separation method. Its equipment is simple, easy to operate and can be used repeatedly without regeneration treatment. Gel chromatography can be used for separation of molecules with different molecular weights. It is widely used in biochemistry, biotechnology and industry, medicine and other fields.

According to Bo Cui, by using Sephadex G-15 column, all the components can be eluted by 2 to 3 column volumes of distilled water. The peak component is confirmed by HPLC analysis to be Mal-β-CD (G2-β-CD) [12].

4.2.3. *Determination of Mal-CDs*

4.2.3.1. *Analysis and detection of Mal-CDs*

The commonly used analysis methods for branched-CD include: paper chromatography (PC), thin layer chromatography (TLC) and HPLC. PC and TLC are simple techniques and easy to operate, require just a little sample and have relatively good resolutions. Thus, PC and TLC are used widely in qualitative and quantitative analysis. First, use PC and TLC for qualitative analysis of Mal-β-CD. Based on

Table 4.2. Comparison of different expension system [12].

Expension system	Volume ratio	β-CD Rf and the effect	Mal-β-CD Rf and the effect
Butanol: Pyridine: Water	6:4:3	0.26 ellipse	0.18 ellipse, tailing (light)
Propanol: Butanol: Water	5:3:4	0.54 ellipse	0.45 ellipse, tailing (light)
75% Isopropanol: EtOH	9:1	0.38 ellipse	0.32 ellipse, tailing (heavy)

the results, establish a quantitative detection of Mal-β-CD by pullulan hydrolysis-Somogyi colorimetry. HPLC can not only faster and more efficiently separate but also more accurately quantify various branched-CDs. The commonly used columns are amino column and octadecyl silica column (ODS). However, the HPLC equipments are expensive. For the economical determination of Mal-CDs, it is needed to compare the results obtained by PC/TLC and those obtained by HPLC.

Filter paper is the supporting material of PC. Different components in a sample have different distribution coefficients and mobile speeds, so as to achieve separation. The rate of solute movement on the filter paper can be expressed by Rf value: Rf = moving distance of solute spot center/moving distance of solvent front. Using Rf value can determine the material qualitatively. Table 4.2 shows the results by several different mobile phases. Mal-β-CD has a relatively large molecular weight. The agent commonly used for separation of oligosaccharides is difficult to expand Mal-β-CD which requires more specific agents and methods. Generally, multiple times expansion and increase of the agent temperature can improve the separation effects. Kobayashi *et al.* obtained a better separation of branched-CD at 70°C [15]. After the color reaction using 1% ethanol, β-CD turns brown, Mal-β-CD is pale yellow, and maltose shows a yellow belt in front of solvent. Because maltose molecules are small and their polarities are high, after multiple times expansion, maltose will be taken to the solvent front.

The operations of TLC and PC are similar, but TLC has more advantages than PC. TLC is faster, easier to separate the mixture and has higher resolution than PC (10–100 times). It can not only separate as low as 0.01 μg of the small samples, but also separate 500 mg or more samples for analysis or preparation.

As shown in Fig. 4.8, β-CD and Mal-β-CD can be well separated by TLC. β-CD exhibits a yellow-brown spot, Mal-β-CD exhibits two light yellow spots

1. Maltose; 2. β-CD standard; 3. Mal-β-CD standard; 4. Sample

Fig. 4.8. Thin-layer chromatogram. Conditions: TLC plate: spread the mixture of silica gel GF254 (5 g) and 0.5% CMC solution (14 mL) onto the 10×20 cm glass plate; concentration of standards (Mal-β-CD, maltose and β-CD) and sample aqueous solution: 10 mg/mL; spotting amount: 10 μL; developer: *n*-propanol: ethyl acetate: water: 25% ammonia = 6:2:5:3; expend twice at 30°C upwardly; dry, irradiate at 254 nm UV light; develop the color reaction by iodine vapor [12].

among which the front one is Mal-β-CD (Rf 0.52), the back one is Mal2-β-CD (*Rf* 0.45) and because of the amount difference the former one is relatively lighter and the latter one is relatively darker. A standard curve can be drawn by analyzing the reducing sugar content in Mal-β-CD spot of each Mal-β-CD standard with different concentration.

Mal-β-CD also can be determined by HPLC as shown in Fig. 4.9(a) and (b).

The concentration of maltose, β-CD and Mal-β-CD standard is 0.4%, 0.3% and 0.3%, respectively. The correction factor can be calculated based on standards' HPLC peak area (maltose 44.76%, β-CD 32.04% and Mal-β-CD 23.20%) and the mass percent of each standard. The correction factor of maltose is calculated to be 0.89, that of β-CD as 0.94 and that of Mal-β-CD as 1.29. Using this method, it can

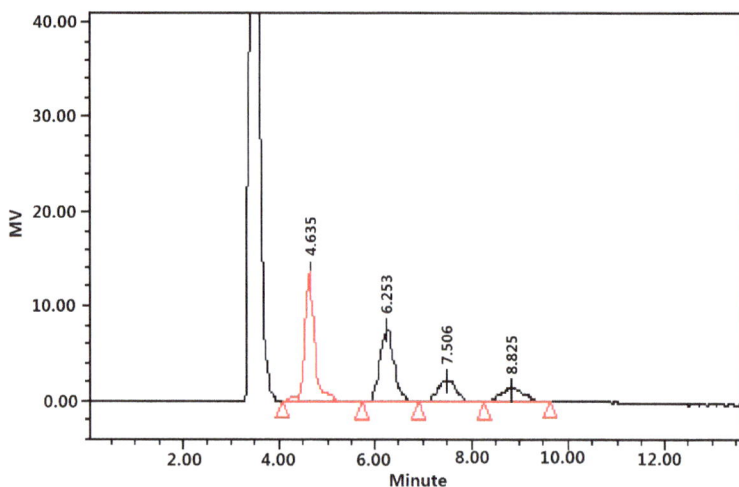

Fig. 4.9(a). High performance liquid chromatograms of the standards. From right to left: Mal2-β-CD, Mal-β-CD, β-CD and maltose. HPLC: waters 600. Column (Hyoersil NH2 5 μm, 4.6 × 250 mm) is maintained at 30°C. Detection: differential refractometer detector (waters 2410). The elution is the mixture of acetonitrile and water (70:30, v/v) at a flow rate of 1.0 mL/min. Injection volume: 10 μL [12].

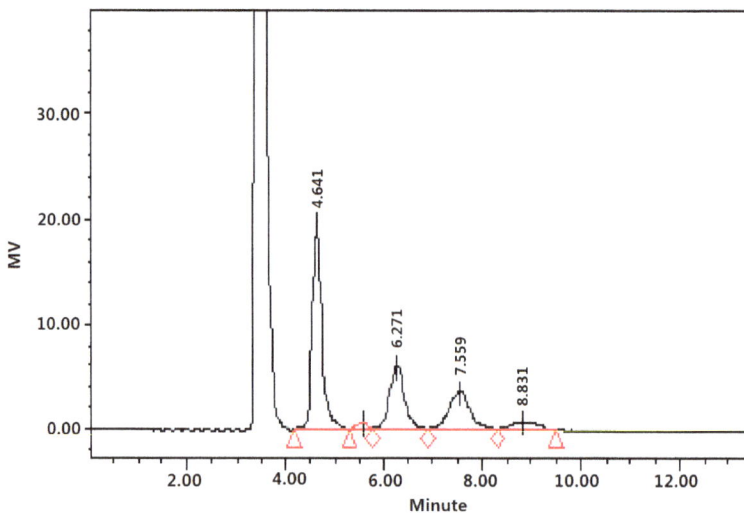

Fig. 4.9(b). High performance liquid chromatograms of the crude Mal-β-CD product. From right to left: Mal2-β-CD, Mal-β-CD, β-CD, branched-tetrose and maltose. HPLC (waters 600). The column (Hyoersil NH2 5 μm, 4.6 × 250 mm) is maintained at 30°C. Detection: differential refractometer detector (waters 2410). The elution is the mixture of acetonitrile and water (70:30, v/v) at a flow rate of 1.0 mL/min. Injection volume: 10 μL [12].

be calculated that the content of Mal-β-CD in the target samples is in agreement with the value calculated by TLC method.

4.2.3.2. Structure identification

Mal-β-CD structure can be analyzed by infrared (IR) spectroscopy, mass spectrometry (MS) and nuclear magnetic resonance (NMR).

There is no significant difference between IR spectra of Mal-β-CD and that of β-CDs with the characteristic peaks of 3414 cm and 2928 cm^{-1} [12]. The peak around 1156 cm^{-1} and 1079 cm^{-1} is because of C–O–C stretching vibration of pyranose ring ether and C–O–H stretching vibration of pyran ring. Peak around 833 cm^{-1} is C–H bending vibration of the α-pyranose, indicating that all the components are α configuration. Peaks around 945, 703 and 576 cm^{-1} are because of CD skeleton vibration. The common feature of Mal-β-CD and β-CD structure is very similar, because Mal-β-CD still has the CD ring structure and is a β-CD derivative.

In the electrospray ionization mass spectrometry (ESI-MS) spectra, separated Mal-β-CD can form two adduct ions $[M + Na]^+$ and $[M + NH4]^+$ and the quasi-molecular ion $[M + H]^+$ with Na^+, $NH4^+$ and H^+, respectively. The m/z is 1482.2, 1477.1 and 1460.0, respectively, and the molecular weight is 1,459 which is in agreement with the theoretical molecular weight value of Mal-β-CD [12].

In the ^{13}C-NMR spectrum, chemical shift δ of C-1 in glucose residues (A–G) of the main ring of CD (Fig. 4.10) is 101.28–101.43, that of C-1 which is connected with H in the α-1, 6 linkages is 98.12, that of C-1 connected with H in the α-1, 4 linkages is 99.45. The chemical shifts δ of glucose residues (A–G) and C-4 of H are 80.60–81.14 and 77.05, respectively; C-4 of sugar residues I, is far away from the main ring, and its chemical shift δ is 68.79. Chemical shift δ of C-6 in the glucose residues A is 66.70. When compared with the chemical shift of C-6 in other glucose residues (59.85), a conclusion can be drawn that C-6 in the glucose residues A has relatively moved which indicates that the side chain is connected on the glucose residue A. In the H NMR spectra, chemical shifts δ of H-1 in the glucose residues (A–G), glucose residue H and glucose residue I is 5.06, 4.94 and 5.36, respectively with an integral intensity ratio of 7:1:1, which indicates that molar ratio of three kinds of connections is 7:1:1.

In heteronuclear multiple-bond correlation spectroscopy (HMBC) spectra, C-1 of glucose residue A–G and the other H-4 generate a coupling peak (δ 101.39, 3.58), and C-4 in a glucose residue A–G and H-1 in another residue produce a coupling peak (δ 80.79, 5.06), which indicates that C-1 and H-4 are connected, and C-4 and H-1 are connected, in accordance with the structure of CD main

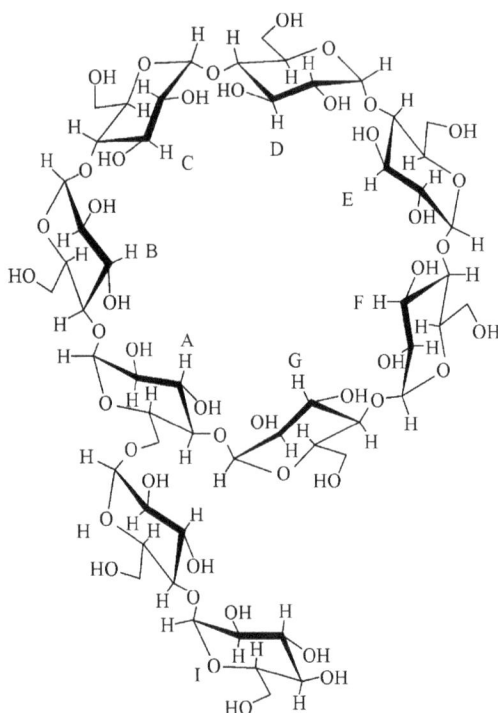

Fig. 4.10.   Molecular structure of Mal-β-CD obtained by NMR analysis [12].

ring where the glucose residue connects end-to-end by α-1, 4 linkages. C-6 in glucose residue A and H-1 in other glucose residue generate a coupling peak (δ 66.70, 4.94). It indicates that C-6 in glucose residue A and H-1 in glucose residue H are connected, in accordance with the structure of mal-β-CD where α-1, 6 glycosidic bond linked residues A and H. Coupling peak (δ 77.05, 5.36) shows that C-4 in glucose residue H is connected to H1 in glucose residue I, which also is in accordance with Mal-β-CD structure where glucose residue H and I are connected.

4.3. Preparation of Gal-CDs

Alpha-D-galactosidase (α-Gal; EC 3.2.1.22) can directly transfer galactosyl residues to the CD rings. For example, add α-Gal to the buffer solution containing melibiose and CDs (α-CD, β-CD, and γ-CD) and incubate at 40°C for 48 h. The reactant contains mixtures of Gal-CDs (Gal-α-CD, Gal-β-CD and Gal-γ-CD), melibiose hydrolyzate, melibiose and CDs. This section will discuss preparation and purification of α-Gal, production of Gal-CDs by α-Gal and Gal-CDs structure analysis.

4.3.1. *Preparation of α-Gal*

α-Gal has been reported to occur widely in a variety of sources: plants, animal tissues and microorganisms [17–20]. This enzyme catalyzes cleavage of the terminal galactosyl residues from a wide range of substrates, including linear and branched oligosaccharides, polysaccharides and various synthetic substrates, such as p-nitrophenyl-α-D-galactopyranoside (PNPG) [21]. It is important for germinating the seeds, and the germination is restrained by deficiency of α-Gal [22]. In the human body, the deficiency of α-Gal leads to Anderson–Fabry's disease [23, 24]. Dormant coffee beans (DCB) became a focus of research for many years, because α-Gal from DCB had two functions: the blood conversion (B-type cell to O-type cell) [25, 26], and the galactosylation of CDs [8]. These effects of α-Gal from DCB were better than that from all other sources on these functions.

4.3.1.1. *Source of α-Gal*

Commercial α-Gal usually originates from two sources: green coffee beans (namely DCB) and *A. niger*. It is difficult to remove the impurities (β-N-acetylglucosaminidase, β-xylosidase and β-galactosidase) during the extraction of α-Gal from *A. niger*. Therefore the main material used to prepare α-Gal is DCB. There have been many studies on applications of α-Gal from DCB, covering wide fields, but the enzyme activity and the yield cannot meet the requirements of application. Koizumi *et al.* [27] reported studies on branched-CDs produced by transgalactosylation with raw α-Gal from DCB, and the enzyme activity of raw α-Gal was found to be low. This restricted the application of α-Gal, and led to α-Gal being expensive in the international market [18, 28]. For obtaining a high yield, Zhu and Goldstein [29] studied cloning and functional expression of a cDNA encoding coffee bean α-Gal. Zhu *et al.* [30] reported high-level expression in the yeast, *Pichia pastoris* and purification of α-Gal. Wangyang Shen *et al.* [31–33] found that during the germination of coffee beans, the activity of α-Gal was different from that in DCB. Moreover, it was higher than the latter in some germinating periods [34]. So they developed a new method to prepare α-Gal from germinating coffee beans (GCB).

4.3.1.2. *Preparation of raw α-Gal from germinated coffee beans*

Before germination, the coffee beans should be soaked at room temperature for 24 h and then spread evenly onto the sand culture (river sand, passed 1 mm sieve, and 3 cm in thickness), covered by a sand layer, and be showered well. Germination can be performed at 26°C, 75% relative humidity and illuminance of approximately 300 LX. Using this method, better survival rate for the germinated coffee beans

Table 4.3. α-Gal enzyme activity and specific activity under different germination times [35].

Parameters	Germination time (d)								
	5	10	15	20	25	30	35	40	45
Activity (nkat)	3.25	5.2	9.6	10.81	12.61	56.94	84.37	70.01	44.81
Protein (mg)	6.79	7.86	10.21	10.88	11.71	16.18	18.62	20.65	26.67
Specific activity (nkat/mg)	0.48	0.66	0.94	0.99	1.08	3.52	4.53	3.39	1.68

can be achieved, as compared with other cultivation methods like sterile culture in tubes. The germinating period is expressed as days in culture (DIC) [34]. Germination time for preparation of α-Gal depends on the internal structural components of the coffee beans, which involves changes in the components during germination where α-Gal content and activity in DCB are the most important. Activity of α-gal in the coffee beans changes with the germination time. Table 4.3 shows the enzyme activity and specific activity under different cultivation periods. The highest enzyme activity and specific activity can be obtained after 35 d cultivation. The GCB should be ground in extraction (sodium chloride, 0.9% w/v, polyvinylpolypyrrolidone (PVPP), 0.9% w/v, ethylene diamine tetraacetic acid (EDTA), 0.1 mM, pH 4.8). Raw α-Gal extracts can be obtained after stirring the above mixture at 4°C for 30 min and centrifuging at 4°C for 15 min at 21,500 g.

Wangyang Sheng optimized the extraction conditions for α-Gal by response surface analysis [35]. And the optimal conditions are: liquid ratio of 1:5, extraction time of 35 min, pH of 5.0. Under these conditions, the total enzyme activity increased 11.6%, and specific enzyme activity increased 11.2% by optimization.

4.3.1.3. *Purification of raw α-Gal*

With the extension of storage time, raw α-Gal extract darkens which adds difficulties to subsequent purification. The deepening of the color is caused by plant polyphenols (PP) in the extract [32–35]. After ammonium sulfate precipitation and before DEAE anion exchange, the extract cannot smoothly pass microfiltration membrane (MFM, 0.45 μm) because of the PP. After filtering through the membrane, PP can stain the DEAE anion-exchange column and reduce the column efficiency and the effects of separation and purification. Compared with other commonly PP removing agents, such as bovine serum albumin (BSA), gelatin, PVPP and alkaloids (caffeine and berberine), etc, acetone exhibits the best

Table 4.4. Effect of each purification step [33].

Separation steps	Total enzyme activity (nkat)	Protein (mg)	Specific activity (nkat/mg)	Yield (%)	Purification (-fold)
Raw extract	129.05	5.62	8.26	100	1
(NH$_4$)$_2$SO$_4$ precipitation	84.14	6.08	13.84	65	1.68
Ion-exchange chromatography	24.67	0.21	117.48	19	14.22
Gel filtration	11.49	0.06	187.17	9	22.66

effect on removal of PP. After treatment with 50% (v/v) acetone, α-Gal's total activity of the extract is 129.05 nkat and specific activity is 8.26 nkat/mg. Total activity is 87.2% of that before removal of PP (148 nkat), and specific activity increased by 15.36% than the latter (7.16 nkat/mg).

After removal of PP, the resulting α-Gal should be further purified by (NH$_4$)$_2$SO$_4$ precipitation, ion-exchange chromatography and gel filtration, in sequence. Table 4.4 shows the effect of each of the above steps.

After the purification steps, satisfactory purification can be achieved.

4.3.2. *Alpha-galactosyl-β-CD preparation conditions*

Alpha-galactosyl-β-CD (α-Gal-β-CD) can be prepared from β-CD and melibiose, through reverse synthesis catalyzed by α-Gal. α-Gal first hydrolyzed melibiose into glucose and galactose, then transfers galactose to β-CD through the reverse hydrolysis reaction and forms galactose branched-CD. This branched-CD has a high water-solubility and the branch galactose has many recognition receptors in animal tissues. Thus, α-Gal-β-CD has broad application prospects.

4.3.2.1. *Amount of melibiose*

Melibiose is one of the substrates of the synthesis reaction. Melibiose is hydrolyzed to fructose and galactose by the help of α-Gal. During the hydrolysis process, the transfer glycoside synthesis is also carried out. Galactose is connected with β-CD through α-1, 6 glycosidic bonds under the transfer glycosides effect of α-Gal to produce α-Gal-β-CD. Melibiose is a provider of galactose which directly affects the amount of the synthetic products. The maximum concentration of melibiose in water is 0.2 mol/L and this concentration can be used in preparation of Gal-β-CD [35].

Table 4.5.   Solubility of β-CD under various temperatures [35].

	Temperature (°C)								
	20	25	30	35	40	45	50	55	60
Solubility (mg/mL)	16.4	18.8	22.8	28.3	34.9	44.0	52.7	60.5	72.9

4.3.2.2. *Amount of β-CD*

It is well known that β-CD exhibits poor water-solubility. Its solubility is not high at various temperatures, as shown in Table 4.5.

Because α-Gal can be inactivated at temperatures above 60°C, only the data below 60°C is listed. In the initial conditions, the reaction temperature is set at 30°C and the melibiose dosage is 0.2 mol/L. At this condition, β-CD can reach supersaturation when its concentration is 0.2 mol/L. Assuming that melibiose can be completely hydrolyzed and the resulting glycosides can be transferred completely on to β-CD, 0.2 mol/L β-CD is needed correspondingly. As the reaction proceeds, β-CD is consumed gradually. So the β-CD that is not dissolved at the beginning of the reaction will be dissolved in the solution as the reaction continues. Furthermore, compared with 0.2 mol/L of melibiose, 0.2 mol/L β-CD is a sufficient amount [35].

4.3.2.3. *Reaction temperature*

As seen from Fig. 4.11, at a certain temperature range, yield of Gal-β-CD increases with increasing temperature. When the temperature is 40°C and 45°C, the yields are stabilized at about 110%, with no major fluctuations. When the temperature is above 45°C, the yield will decrease rapidly. When the temperature is 60°C, the yield is zero, because the enzyme is inactivated and the too long reaction time (30 h) can lead to enzyme inactivation. Therefore, the proper reaction temperature is 40°C [35].

4.3.2.4. *Shaking speed*

As seen from Fig. 4.12, the yield of Gal-β-CD increases with the increase of shaking speed and levels when the shaking speed is higher than 75 r/min [35].

4.3.2.5. *α-Gal dosage*

The yield of Gal-β-CD increased with the increase of enzyme dosage at the beginning and levels off when the enzyme dosage reached 75 nkat (Fig. 4.13). Further increase of enzyme dosage has no effect on the yield of Gal-β-CD.

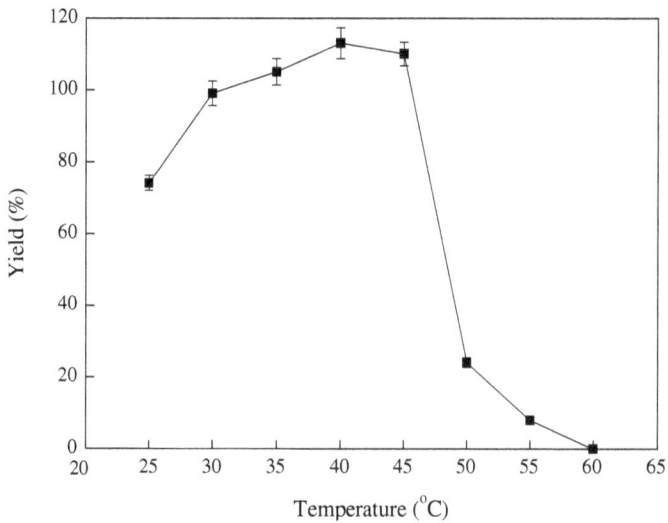

Fig. 4.11. Effect of temperature on the yield of Gal-β-CD. Reaction conditions: β-CD 0.2 mol/L, melibiose 0.2 mol/L, α-Gal 50 nkat, 2 mL acetate buffer (50 mmol/L, pH 6), reaction time 36 h, the shaking speed 100 r/min [35].

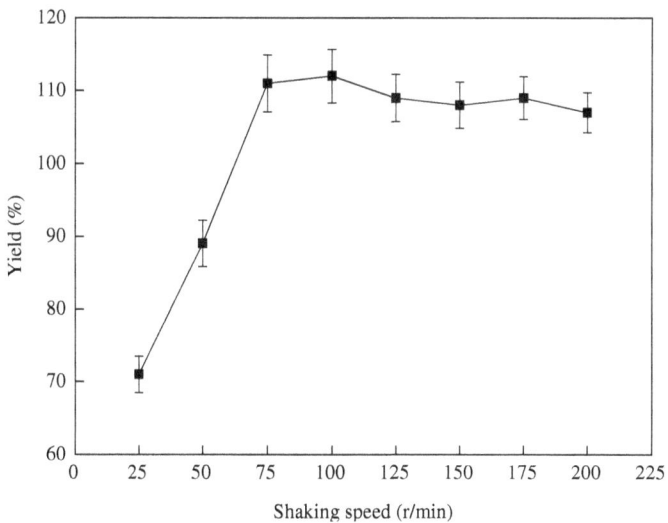

Fig. 4.12. Effect of shaking speed on yield of Gal-β-CD. Reaction conditions: β-CD 0.2 mol/L, melibiose 0.2 mol/L, α-Gal 50 nkat, 2 mL of 50 mmol/L acetate buffer (pH 6), reaction time of 36 h and temperature of 40°C [16].

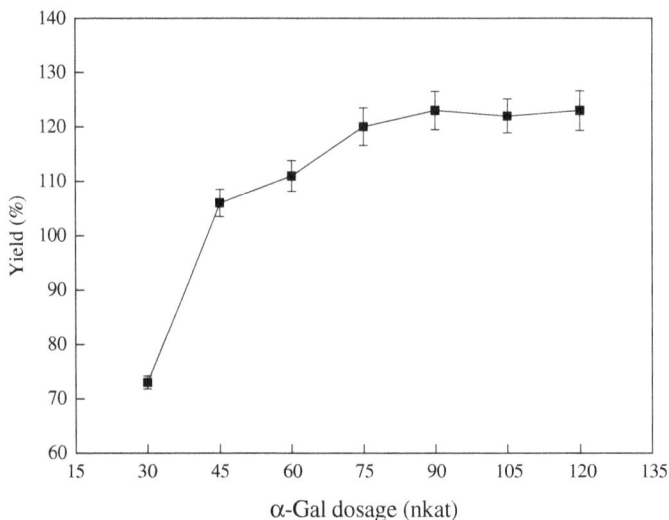

Fig. 4.13. Effect of α-Gal dosage on yield of Gal-β-CD. Reaction conditions: β-CD 0.2 mol/L, melibiose 0.2 mol/L, 2 mL of 50 mmol/L acetate buffer (pH 6), reaction time of 36 h, temperature of 40°C and shaking speed of 75 r/min [35].

4.3.2.6. *pH value*

α-Gal is very sensitive to the pH of the reaction system. The highest yield of Gal-β-CD can be obtained when the system pH is 6.5 (Fig. 4.14). Too acidic and alkaline environments are not suitable for the production of Gal-β-CD because the activity of α-Gal can be inhibited.

4.3.2.7. *Reaction time*

Enzymatic production of Gal-β-CD by α-Gal is a transglycosidation reaction. It is very important to choose a proper time. If the reaction time is too short, the reverse reaction cannot reach to the maximum which will influence the yield of Gal-β-CD. On the contrary, when the reaction time is too long, the produced Gal-β-CD can be hydrolyzed which will result in the drop of the yield. Therefore, the reaction time of control is very important. Figure 4.10 shows the effect of reaction time on the yield of Gal-β-CD.

First, the yield of Gal-β-CD increased with the increase in reaction time and reached the highest after reaction for 24 h. Afterwards the yield of Gal-β-CD sharply decreased (Fig. 4.15). The produced Gal-β-CD is hydrolyzed by prolonged reaction (>24 h).

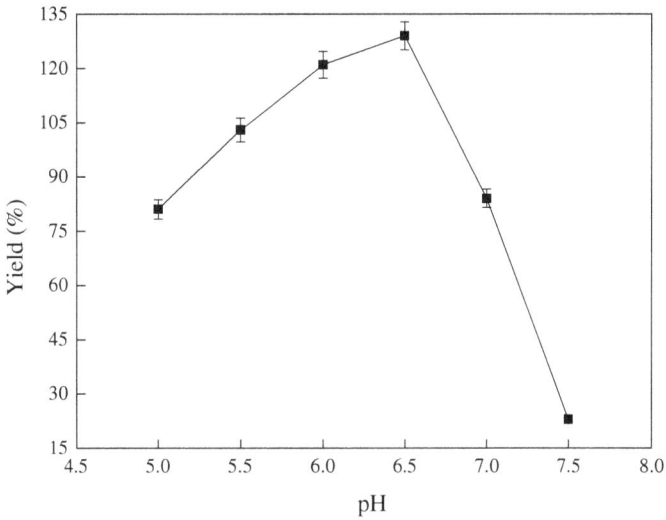

Fig. 4.14. Effect of pH on yield of Gal-β-CD. Reaction conditions: β-CD 0.2 mol/L, melibiose 0.2 mol/L, enzyme dosage of 75 nkat, reaction time of 36 h, temperature of 40°C and shaking speed of 75 r/min [35].

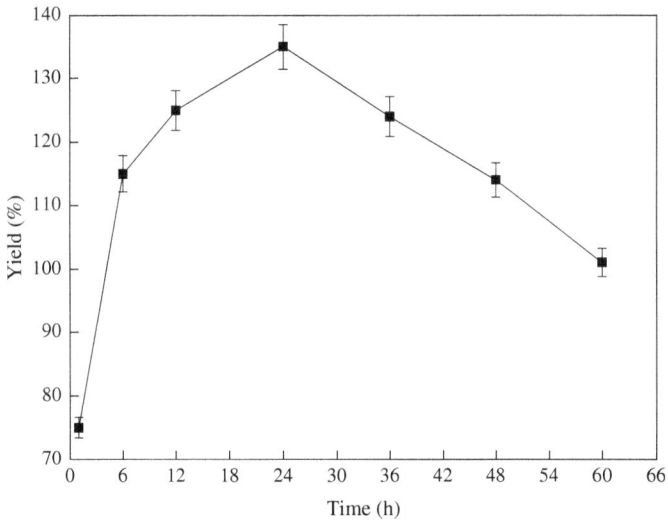

Fig. 4.15. Effect of reaction time on yield of Gal-β-CD. Reaction conditions: β-CD 0.2 mol/L, melibiose 0.2 mol/L, enzyme dosage of 75 nkat, 2 mL of 50 mmol/L acetate buffer (pH 6), temperature of 40°C and shaking speed of 75 r/min [35].

According to Wangyang Shen *et al.*, the optimal conditions for the production of Gal-β-CD obtained by response surface analysis are: β-CD 0.2 mol/L, melibiose 0.2 mol/L, enzyme dosage of 73 nkat, 2 mL of 50 mmol/L acetate buffer (pH 6.3), temperature of 40°C, reaction time of 28 h and shaking speed of 75 r/min [16]. The conversion rate (%) of Gal-β-CD under the optimized conditions is 28% (conversion rate (%) = mole number of Gal-β-CD/mole number of β-CD × 100%).

4.3.3. *Analytical techniques*

4.3.3.1. *Determination of Gal-β-CD by HPLC*

The HPLC analysis conditions for crude Gal-β-CD and β-CD, melibiose and galactose standard are all the same. According to each peak retention time in the HPLC chromatogram of crude Gal-β-CD (Fig. 4.16) and that of standards (figure not shown), the crude product contains β-CD (peak retention time for the standard and that of crude Gal-β-CD is 13.537 min and 14.136 min, respectively), melibiose (peak retention time for the standard and that of crude Gal-β-CD is 7.020 min and 7.517 min, respectively), galactose (peak retention time for the standard and that of crude Gal-β-CD is 5.068 min and 4.841 min, respectively), etc. According to the characteristics of HPLC separation column, the last coming out peak (retention time of 20.402 min) of the crude Gal-β-CD product in the HPLC chromatogram is the typical peak for Gal-β-CD.

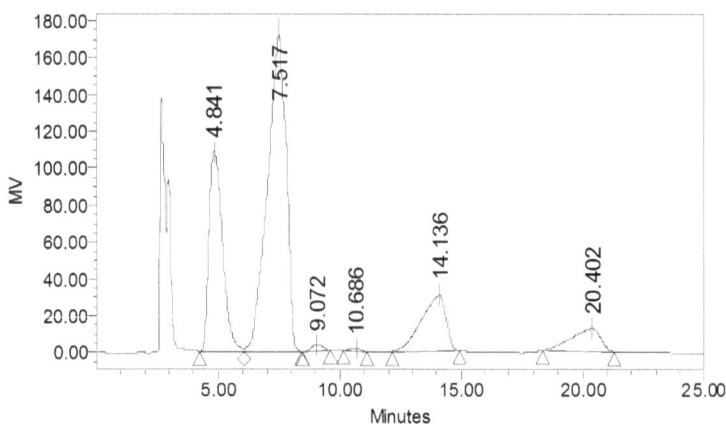

Fig. 4.16. High performance liquid chromatogram of Gal-β-CD product. Column: Hypersil Asp-1-NH2 column (4.6 × 200 mm); mobile phase: 75% acetonitrile, 25% water; detector: waters 2410 refractive index detector; column temperature: 30°C; flow rate: 1 mL/min [35].

Highly purified Gal-β-CD can be obtained by using preparative HPLC. Column of the preparative HPLC is the reversed-phase C18 column. Branched-β-CD (retention time 9.798 min) comes earlier than β-CD (retention time 11.808 min). In order to obtain high purity products, collection begins after the peak comes out, and ends after the peak completely disappears. After preparative HPLC, pure Gal-β-CD powder can be obtained by freeze-drying the purified Gal-β-CD solution.

4.3.3.2. *Structural identification of Gal-β-CD*

4.3.3.2.1. Composition analysis

There are mainly two methods for chemical composition analysis: hydrolysis method and MS method. Hydrolysis method is by adding α-Gal to induce hydrolysis reaction and after the reaction is analyzed by HPLC; MS is the direct detection of the product composition mass/charge ratio and can more accurately determine the composition of the product chemical components.

(i) Hydrolysis method

Take 100 mg of the product powder, dissolve it in 2 mL of 50 mmol/L acetic acid buffers (pH 6), and add sufficient quantities of α-Gal (50 nkat) and stir to dissolve. The reaction temperature is 30°C, reaction time is 36 h and the shaking speed is 100 r/min. After the enzymatic reaction, put the container into boiling water bath for 10 min and the α-Gal can be fully inactivated. Add a certain amount of ultra-pure water to the inactivated mixture, and transfer into the centrifuge tube and centrifuge at 5,000 r/min for 5 min. Take the supernatant for HPLC analysis. The results are shown in Fig. 4.17.

Comparing with the HPLC peak retention time of standard samples, it can be concluded that the main component is galactose (peak retention time of the standard is 5.068 min and that of the product is 4.975 min) and β-CD (peak retention time of the standard is 13.537 min and that of the product is 13.551 min). Quantification of the amount of galactose and β-CD by external standard method obtained that the mole ratio of the two components is 1:1. It can be speculated that the main product is a single-substituted galactose-β-CD derivative.

(ii) MS analysis

Take 10 mg of the product powder, dissolve in 2 mL ultra-pure water and analyze directly. Figure 4.18 is the ESI-MS spectrum of the product.

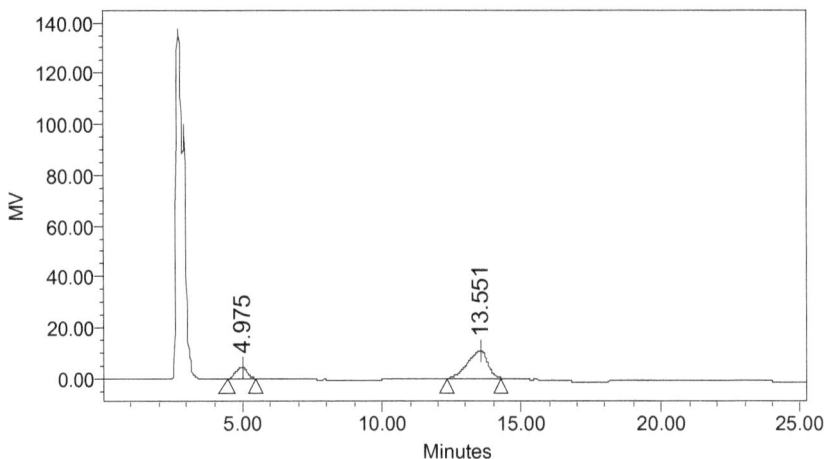

Fig. 4.17. High performance liquid chromatogram of hydrolyzed Gal-β-CD product. Column: Hypersil Asp-1-NH2 column (4.6 × 200 mm); mobile phase: 75% acetonitrile, 25% water; detector: waters 2410 refractive index detector; column temperature: 30°C; flow rate: 1 mL/min [35].

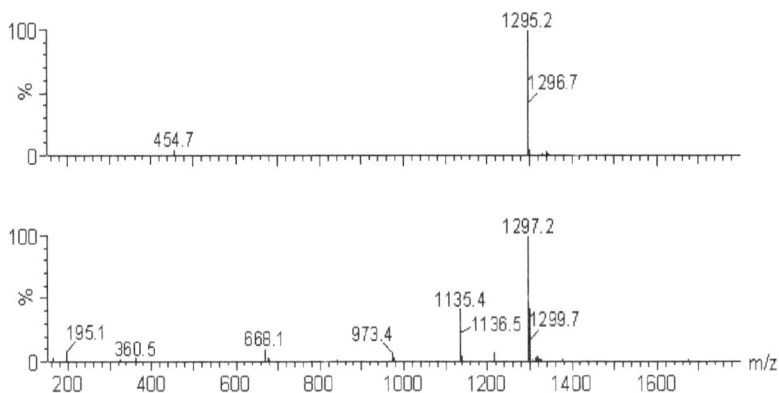

Fig. 4.18. ESI-MS spectrum of the product. Ionization mode: ESI; spray voltage: 3.7 KV; ion source temperature: 120°C; desolvation temperature: 300°C; ion energy: 1.0 v [35].

Figure 4.18 shows that: mass charge ratio (m/z) of the negative ions is 1295.2 and the mass charge ratio (m/z) of the positive ions is 1297.2, which is consistent with the molecular weight of Gal-β-CD [35]. There are rare fragments with other mass charge ratio (m/z) in the positive and negative ion mass spectra, indicating that the main product is single-substituted Gal-β-CD.

4.3.3.2.2. Structure analysis

(i) IR spectroscopy

Take 1 mg sample powder into an agate mortar, mix with about 100 mg potassium bromide and grind into fine powder. Transfer the mixture into the mold and press into test tables for IR spectroscopy [35]. The results are shown in Fig. 4.14.

Both IR spectra of the product and β-CD exhibit the characteristics of carbohydrate absorption, i.e. 3,400 cm^{-1} (O–H stretching), 2,930 cm^{-1} (C–H stretching) and 1,640 cm^{-1} (O–H bending), 1,155 cm^{-1} (C–O vibration), etc. The absorption around 855 cm^{-1} is characteristic of α-glycosidic bond, indicating that the product contains α-glycoside linkage (α-1, 2, α-1, 3, α-1, 4, α-1, 6 linkages) of which α-1, 4 glycosidic bond is of β-CD specified. β-CD consists of seven glucose residues connected end-to-end into a ring structure by α-1, 4 glycosidic bond. The galactosyl may be connected to β-CD through the α-1, 2, α-1, 3 or α-1, 6 glycosidic bond, rather than the α-1, 4 glycosidic bond), and no β-glucoside linkage. The strong absorption peak around 3400 cm^{-1} is the hydroxyl absorption peak. After the formation of an intramolecular hydrogen bond, the bond force constants decrease, resulting in broad and strong absorption. The shape of β-CD peak is wider than that of the product, because introduction of galactose group partly breaks the hydrogen bonds within the CD molecules.

From the product IR spectrum (Fig. 4.19), it can be concluded that in the product the galactose group is attached to β-CD through α- glycosidic bond. The type of glycosidic bonds may be: α-1, 2, α-1, 3 or α-1, 6 glycosidic bonds.

(ii) NMR spectra

^{13}C-NMR spectra of β-CD and product are determined by a Bruker Avance NMR in the frequency of 500 MHz. Dissolve the product powder in D$_2$O at 60°C and transfer into the 5 mm diameter tube for testing. Figure 4.20 is the ^{13}C-NMR spectrum of the product.

According to the characteristics of ^{13}C-NMR and the references [4, 27, 36, 37], the chemical shift and assignment for the carbon atoms in Gal-β-CD can be obtained.

From Fig. 4.20 and Table 4.6, we can get the following message: There are two C-1 signals, i.e. C-1 signal which is connected by α-1, 4 linkage ($\delta \sim 104.43$) and C-1 signal connected by α-1, 6 linkage ($\delta \sim 101.28$), respectively. Their signal intensity ratio is 7:1. There are two kinds of C-4 signals, i.e. C-4 signal on the β-CD glucose residues ($\delta \sim 83.79$) and C-4 signal which is from the galactose residues ($\delta \sim 71.79$), respectively. The intensity ratio is 7:1. There are three C-6 signals. One is C-6 signal on the β-CD glucose residues ($\delta \sim 62.95$), another is the C-6 signal on

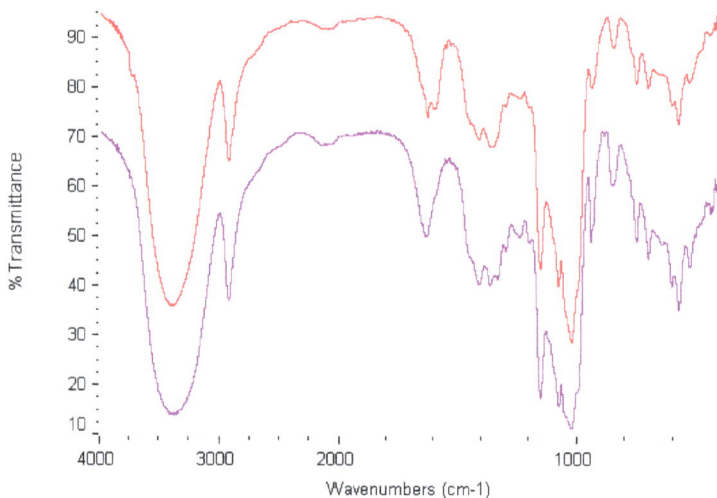

Fig. 4.19. IR spectra for the product (bottom) and β-CD (top). NEXUS fourier transform IR spectrometer; DTGS detector; scanning frequency: 32; resolution: $4\,\mathrm{cm}^{-1}$ [35].

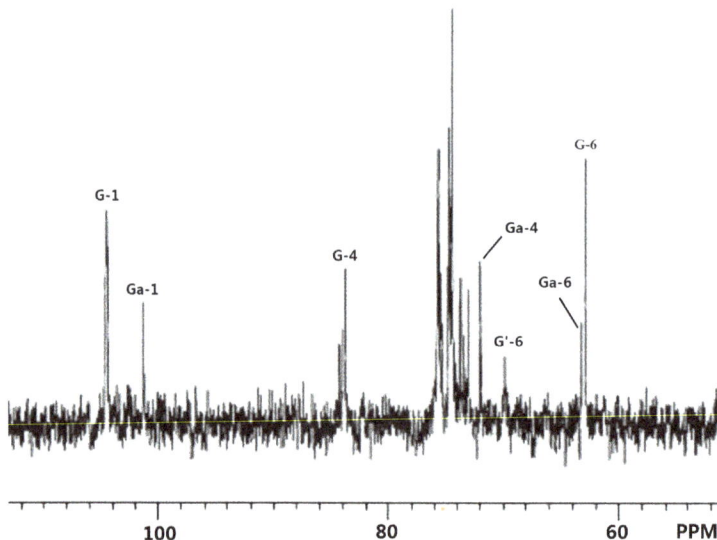

Fig. 4.20. ^{13}C NMR spectra (500 MHz) of the product. G represents glucose residues; Ga represents galactose residues; G′ glucose residue which is connected with a galactose residue; numbers represent the number of carbon atoms in the glucose residues [35].

Table 4.6. Chemical shift and assignment of the carbon atoms in Gal-β-CD [35].

Position	Number					
	C-1	C-2	C-3	C-4	C-5	C-6
G	104.43	74.70	75.69	83.79	74.45	62.95
G'	⋆	⋆	⋆	⋆	⋆	69.65
Ga	101.28	73.74	⋆	71.79	73.03	63.17

⋆uncertain

the galactose residues ($\delta \sim 63.17$) and the other C-6 signal ($\delta \sim 69.65$) had greater downfield shift, indicating that the side chain galactose residues is connected with the oxygen of these carbon atoms. The intensity ratio of these three C-6 signals is 6:1:1. According to ^{13}C-NMR, it can be concluded that in the product the galactose group is attached to β-CD through α-1, 6 glycosidic bonds.

Based on the results of hydrolysis analysis, MS, IR spectroscopy and ^{13}C-NMR analysis, the product can be confirmed to be single-substituted Gal-β-CD with galactose group and is attached to β-CD through α-1, 6 glycosidic bonds. Its schematic diagram is shown in Fig. 4.21.

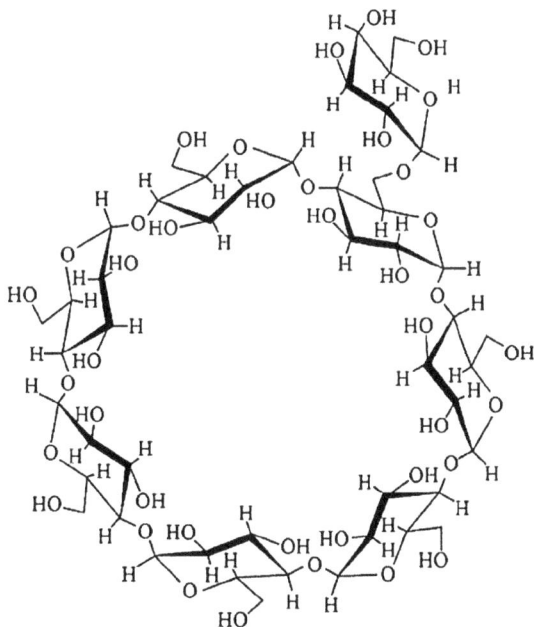

Fig. 4.21. Fine structure of Gal-β-CD [35].

4.4. Other Enzyme Modified CD

4.4.1. *Glucosyl-CD*

Glucosyl-α-(1,6)-CDs (G1-CD), i.e. one glucose attached to C-6 of the CDs pyranose units through α-1, 6 glycosidic linkage, is a kind of CD derivative. Preparation of G1-CD is usually started with starch and CGTase. After CDs are generated, put glucoamylase into the reaction system, and then one mole glucose can attach onto one CD ring. In the presence of surfactant SDS, branched-CDs with different degree of polymerization can be prepared under the reaction of CGTase [15]. G1-CDs also can be obtained by the combined reaction of amyloglucosidase and *Aspergillus oryzae* α-amylase. The product is roughly composed of G1-α-CD (26%), G1-β-CD (11.5%), G1-γ-CD (6.6%), α-CD (13%), β-CD (3.5%) and glucose (39.4%). However, the latter method results in purification problems and low yield.

4.4.2. *Mannosyl-glucosyl-CD (Man-G1-CD)*

α-mannosidase can be extracted from jack beans and almonds [37]. The optimal pH of the jack bean α-mannosidase is 4.5 and that of almond α-mannosidase is 6.0. Add methyl-α-mannopyranoside and branched-CDs (G1-α-CD, G1-β-CD, G2-α-CD or G2-β-CD) to the proper buffer (depends on the enzyme source) at 40°C and incubate for 15 h. Use HPLC for product separation and ^{13}C-NMR for product identification. There are four kinds of Man-G1-CDs: α-D-mannosyl-(1,6)-α-D-glucosyl-(1,6)-α-CD, α-D-mannosyl-(1,6)-α-D-glucosyl-(1,6)-β-CD, α-D-mannosyl-(1,6)-α-D-glucosyl (1,4)-α-D-glucosyl-(1,6)-α-CD and α-D-mannosyl-(1,6)-α-D-glucosyl (1,4)-α-D-glucosyl-(1,6)-β-CD.

4.4.3. *Galactosyl-glucosyl-CD*

Galactosyl-glucosyl-CDs (Gal-G1-CDs) can be prepared by using branched-CDs, such as Gl-α-CD or G1-β-CD, as receptors, using galactose as donor and using coffee bean α-Gal to transfer galactose [1]. Separation of the products by HPLC shows that the product can be hydrolyzed to equimolar galactose and G1-CD; FAB-MS data show that the resulting three kinds of products, namely, Gal–Glc group on the C-6 position, Gal and Glc groups connect to C-6 positions of two glucopyranoses, respectively, and Gal and Glc groups connect to C-6 position of one glucopyranose and C-2 (or C-3) position of one glucopyranose, respectively.

4.4.4. *Galactosyl-maltosyl-CD*

Using Mal-Cds as the receptor, galactose as donor and coffee bean α-Gal to transfer galactose, three kinds of Galactosyl-maltosyl-CD (Gal-Mal-CD) products can be obtained [1]. FAB-MS data show that the structures are: galactose and maltose groups directly bond onto the glucopyranose of CDs ring and the galactose group is connected to maltose side chain.

References

1. Tong, LH (2001). *Cyclodextrin Chemistry–Foundation and Application.* Beijing: Science Press.
2. Ajisaka, N, K Hara, K Mikuni and H Hashimoto (2000). Effects of branched cyclodextrins on the solubility and stability of terpenes. *Bioscience, Biotechnology, and Biochemistry*, 64(4), 731–734.
3. Shoichi, K, N Kohichi and A Masaomi (1989). Production and some properties of branched cyclomalto-oligosaccharides. *Carbohydrate Research*, 192, 223–231.
4. Kitahata, S, Y Yoshimura and S Okada (1987). Formation of 6-O-α-maltosyl cyclo-maltooligosaccharides from α-maltosyl fluoride and cyclomaltooligosaccharides by pullulanase. *Carbohydrate Research*, 159(2), 303–313.
5. Shen, WY and ZY Jin (2005). Production of 6-O-α-D-maltosylcyclomaltoheptaose by reverse reaction of thermostable bacillus acldopullulyticus pullulanase. *Science and Technology of Food Industry* (*Chinese*), 26(9), 105–107.
6. Cui, B and ZY Jin (2005). Synthesis of maltosyl (α-1→6)β-cyclodextrin through the reverse reaction of pullulanase. *Food science* (*Chinese*), 26(12), 128–131.
7. Cui, B and ZY Jin (2005). Quantitative determination of maltosyl (α-1→6)β-cyclodextrins with paper chromatography. *Journal of Food Science and Biotechnology*, 24(6), 88–91.
8. Cui, B and ZY Jin (2007). Enzymatic synthesis and identification of maltosyl (α-1→6) β-cyclodextrin. *Chemical Research in Chinese Universities* (*Chinese*), 28(2), 283–285.
9. Noriyasu, W (1997). A novel method to produce branched α-cyclodextrins: Pullulanase glucoamylase mixed method. *Journal of Fermentation and Bioengineering*, 83(1), 43–47.
10. Hizukiri, S, S Kawano and J Abe *et al.* (1989). Production of branched cycloaml-tooctaose through the reverse action of *Klehsiella aerogenes* pullulanase. *Biotechnology Application Biochemistry*, 11, 60–73.
11. Wang, SJ and ZY Jin (2005). Production of 6-O-α-D-maltosylcyclomaltoheptaose by reverse reaction of thermostable bacillus acldopullulyticus pullulanase. *Science and Technology of Food Industry* (*Chinese*), 26(9), 105–107.
12. Cui, B (2011). Study on enzymatic synthesis characterisitics and application of maltosyl (α-1→6)β-cyclodextrin. Doctor's degree Thesis, School of Food Science and Technology. Wuxi: Jiangnan University.
13. Sarney, DB, C Hale, G Frankel and EN Vulfson (2000). A novel approach to the recovery of biologically active oligosaccharides from milk using a combination of enzymatic treatment and nanofiltration. *Biotechnology and Bioengineering*, 69(4), 461–467.

14. Aydogan, N, T Gurkan and L Yilmaz (1998). Effect of operating parameters on the separation of sugars by nanofiltration. *Science and Technology*, 33, 1767–1785.
15. Whistler, RL and DG Durso (1950). Chromatographic separation of sugars on charcoal. *Journal of the American Chemical Society*, 72, 677–679.
16. Kobayashi, S, N Shibuya, BM Young and D French (1984). The preparation of 6-O-α-d-glucopyranosylcyclohexaamylose. *Carbohydrate Research*, 126(2), 215–224.
17. Barham, D, PM Dey, D Griffiths and JB Pridham (1971). Studies on the distribution of α-galactosidases in seeds. *Phytochemistry*, 10(8), 1759–1763.
18. Carchon, H and CK DeBruyne (1975). Purification and properties of coffee-bean alpha-D-galactosidase. *Carbohydrate Research*, 41, 175–189.
19. Dey, PM and JB Pridham (1972). Biochemistry of α-galactosidase *Advances in Enzymology and Related Areas of Molecular Biology*, 72(36), 91–130.
20. Skленá Ťova, S and M Tichá (1991). Electrophoretic study of α-D-mannosidase and α-D-galactosidase from dry seeds of *Pisum sativum*. *Journal of Chromatography A*, 540, 365–372.
21. Simerska, P, M Kuzma, D Monti, S Riva, M Mackova and V Kren (2006). Unique transglycosylation potential of extracellular alpha-D-galactosidase from *Talaromyces flavus*. *Journal of Molecular Catalysis B-Enzymatic*, 39(1–4), 128–134.
22. Mathew, CD and K Balasubramaniam (1986). Chemical modification of α-galactosidase from coconut. *Phytochemistry*, 25, 2439–2443.
23. Peters, FPJ, A Vermeulen and TL Kho (2001). Anderson-Fabry's disease: Alpha-galactosidase deficiency. *Lancet*, 357(9250), 138–140.
24. Takenaka, T, G Qin, RO Brady and JA Medin (1999). Circulating alpha-galactosidase A derived from transduced bone marrow cells: Relevance for corrective gene transfer for Fabry disease. *Human gene therapy*, 10(12), 1931–1939.
25. Lenny, LL and J Goldstein (1991). The production of group O cells. *Biotechnology*, 19, 75–100.
26. Lenny, LL, R Hurst, J Goldstein, LJ Benjamin and RL Jones (1991). Single-unit transfusions of RBC enzymatically converted from group B to group O to A and O normal volunteers. *Blood*, 77(6), 1383-1388.
27. Koizumi, K, T Tanimoto, K Fujita, K Hara, N Kuwahara and S Kitahata (1993). Preparation, isolation, and characterization of novel heterogeneous branched cyclomalto-oligosaccharides having β-d-galactosyl residue(s) on the side chain. *Carbohydrate Research*, 238(0), 75–91.
28. Harpaz, N, HM Flowers and N Sharon (1974). Purification of coffee bean alpha-galactosidase by affinity chromatography. *Biochimica et biophysica acta*, 341(1), 213–321.
29. Zhu, A and J Goldstein (1994). Cloning and functional expression of a cDNA encoding coffee bean alpha-galactosidase. *Gene*, 140(2), 227–231.
30. Zhu, A, C Monahan, Z Zhang, R Hurst, L Leng and J Goldstein (1995). High-level expression and purification of coffee bean alpha-galactosidase produced in the yeast *Pichia pastoris*. *Archives of biochemistry and biophysics*, 324(1), 65–70.
31. Shen, WY and ZY Jin (2008). Effect of plant polyphenols on extraction of α-D-galactosidase. *Science and Technology of Food Industry* (*Chinese*), 29(4), 90–92.

32. Shen, WY and ZY Jin (2008). Study on the property of α-galactosidase from the germinating coffee bean. *Journal of Anhui Agricultural Sciences* (*Chinese*), 36(5), 1770–1772.

33. Shen, WY, ZY Jin, XM Xu, JW Zhao, L Deng, HQ Chen, C Yuan, DD Li and XH Li (2008). New source of alpha-D-galactosidase: Germinating coffee beans. *Food Chemistry*, 110(4), 962–966.

34. Marraccini, P, WJ Rogers, V Caillet, A Deshayes, D Granato, F Lausanne, S Lechat, D Pridmore and V Petiard (2005). Biochemical and molecular characterization of alpha-D-galactosidase from coffee beans. *Plant Physiology and Biochemistry*, 43(10–11), 909–920.

35. Shen, WY (2009). Preparation, properties and application of α-galactosidase from germinating coffee beans. Ph. D thesis. Wuxi: Jiangnan University.

36. Okada, Y, K Koizumi and S Kitahata (1996). Isolation by HPLC of the positional isomers of heterogeneous doubly branched cyclomaltohexaose having one α-d-galactosyl and one α-d-glucosyl side chain and determination of their structures by enzymatic degradation. *Carbohydrate Research*, 287(2), 213–223.

37. Hara, K, K Fujita, H Nakano, N Kuwahara, T Tanimoto, H Hashimoto, K Koizumi and S Kitahara (1994). Acceptor specificities of α-Mannosidases from jack bean and almond, and trans-mannosylation of branched cyclodextrins. *Bioscience Biotechnology and Biochemistry*, 58(1), 60–63.

5

PREPARATION AND ANALYSIS OF CYCLODEXTRIN DERIVATIVES

Chao Yuan*, Yu-Xiang Bai[†] and Zheng-Yu Jin[‡]

**School of Biotechnology and Food Science, Anyang Institute of Technology
Anyang 455000, China
[†]School of Food Science and Technology, Jiangnan University
Wuxi 214122, China
[‡]The State Key Laboratory of Food Science and Technology,
School of Food Science and Technology, Jiangnan University,
Wuxi 214122, China*

As the native cyclodextrins (CDs) exhibit some limitations in application, they need to be modified to improve the properties. All the modification methods could be divided into two kinds, chemical modification and enzymatic modification. Based on the stable cyclic structure, CDs could be modified via etherification, esterification, oxidation and crosslinking reactions. The chemical modification has a special purpose of introducing novel functional group. CDs modified by chemical means were named as cyclodextrin chemical derivatives (CCDs).

In this chapter, synthesis pathway, separation and detection of several important CCDs will be discussed.

5.1. The Basics of CCDs Preparation

5.1.1. *Chemical method and pathway of CD modification*

Chemical modification was first investigated. Researchers began to study the modified CD to expand the applications of native CDs. Although CDs contained a cavity to include the guest molecule, they had difficulties to form functional super molecules. Thanks to the stability, CDs could be tridimensionally modified. Because of the needs in supramolecular applications, a lot of work had been done on chemically modified CDs. Specially, several papers on CCDs had been published during the recent 20 years [1]. However, a majority of the papers focused on the application but not the synthesis. Different categories of CDs were named by varied definitions given by different researchers. In this chapter, the common classification is adopted.

All the chemical modification depends on the structure of basic unit D-glucose and the chemical characters of bond. Most of the CCDs were prepared by reacting with the hydroxyl in C-2, C-3, C-6 or other kinds of bonds, including C–O and C–H [2].

(1) Hydroxyl (–OH)

Almost all the CDs are modified through chemical reactions on the hydroxyl groups. The reactivity of the three kinds of hydroxyls follows the order: C-6>C-2 >C-3. The acidities sequence is C-2>C-3≫C-6. The C-6 hydroxyl is most easily replaced via the reaction of CDs with reagent in a strong alkaline environment. However, in the weak alkaline environment, the C-2 hydroxyl is easier to be activated and replaced. Because of the steric hindrance, the large molecular size reagent is more convenient to attack the C-6 hydroxyl. The C-3 hydroxyl can be replaced only if the C-2 and C-6 hydroxyls are shielded. When the CDs are used to encapsulate the guest, the dimension of the cavity plays the key role in the stability of the complexes. But the main factor of unstable complexes is nucleophilicity. In addition, the reaction medium and cavity size both had effect on the structure of the reaction product.

Hydroxyl could always react with alkyl halide, epoxide and a series of halides to obtain the ester, ether and even the polymers or CCDs. An electron withdraw group is needed to activate the oxygen atom while the C–OH bond crashes.

(2) C–O–C bond

The CDs are stable when treated by acid in general. But the C–O–C bond also could be broken by lewis acid mixed with the reductive agent. This kind of reaction would not disturb the substituted group of glucose unit. The special phenomenon can be applied to analyze the distribution of all substituted groups. CDs can be hydrolyzed into glucose unit and then detected by chromatographic method to confirm the number and position of the substituted group.

(3) 2, 3 C–C bond

2, 3 C–C bond can be oxidatively cleaved by periodate or iodate to obtain aldehyde group macrocycle compounds. Then aldehyde group can be further reduced to alcohol, or involved in other reactions.

There is confusion in the characterization of the degree of substitution (DS) of the relevant modified CDs. Some researchers agreed that the DS represented the number of the substituted groups located in each glucose unit. Thus, the DS value is between 0–3. Whereas some others agreed that the DS was based on the number of substitute groups on each CD unit. In this case the DS value could be more than three.

Various groups can be introduced to CDs by the above reactions. However, because the α-, β-and γ-CD molecules have a similar structure formed by 6, 7 and 8 glucose units, and with almost no difference of the 2-, 3-, 6-hydroxyl groups among all the glucose units, it is hard to know how the substitution happens and how to separate the products. In order to facilitate positioning of the distribution of substituents, the glucose unit of the CDs are named as A, B, C ... Generally, the glucose unit which had a substituent could be ranked as A, followed by other numbers.

According to the related chemical reaction theory on chemically modified CDs, we adopt the following method to control the substitution position.

(1) Control of the reaction conditions

Reagents could react with different hydroxyls in different solvents. A typical example is the preparation of hydroxypropyl CD. In weak alkaline conditions, propylene oxide prefers to replace the 2-hydroxyl, 3-hydroxyl, but replaces the 6-hydroxyl in strong alkaline solution. This phenomenon is caused by reactivity and steric hindrance of the three free hydroxyl groups. In the selective modification, the substituents could partly replace the specific hydroxyl through the regulation of pH, modifiers, solvents (water, water–ethanol) and reaction time.

(2) Intermediate positioning

For example: when sulfonation reagent *p*-toluenesulfonyl chloride (TsCl) reacts with the CD under appropriate conditions [3], 6 or 2, 3 single substituted CDs would be generated. Then these substituted CDs can be treated by nucleophiles in order to obtain the deoxy derivatives. Specific disubstitution modified CDs are shown in Table 5.1 and the corresponding structure of the bridge group are shown in Fig. 5.1.

Table 5.1.　Bridge linking disubstitution modified CDs [3].

Order	CD	Bridge group	Link mode
1	β	–Ph–	6^A6^B
2	β	–Ph(MeO)$_2$–	6^A6^B
3	β	–Ph–Ph–	6^A6^D
4	β	–Ph–CO–Ph–	6^A6^C
5	β	–Ph–CHCH–Ph–	6^A6^D
6	β	9,10-dicyanoanthraquinone	6^A6^D

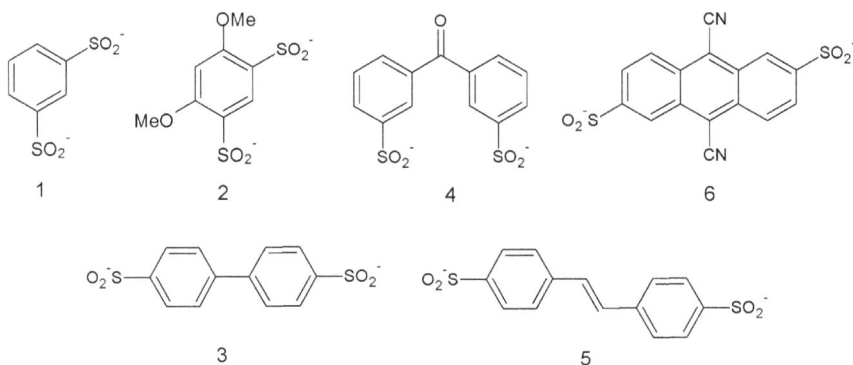

Fig. 5.1.　Structure of bridge groups [3].

(3) Selective positioning by blocking groups

In order to selectively modify the 6-hydroxy, the 2, 3-hydroxy, specially the 2-, can be shielded by binding with blocking groups. They reacts with the modification reagent to achieve the modified CDs. For example, first replace all the three hydroxyls of glucose unit using sulfonyl reagent, then remove the 6- substituent by 2- propanol potassium which is mixed in 2-propanol/benzene solution. Thereafter, the relevant modifiers can react with 6- only. Finally, remove the 2, 3- benzoyl via methanol/potassium hydroxide solution.

(4) Breakage of glycoside bond

It is reported that acetylation or methylation CD treated by PhsTMs–ZnI$_2$ can cleave only one C–C bond and generate high yield of malto-oligosaccharide. The internal cyclization would occur on this modified malto-oligosaccharides to produce new single-modified CDs.

5.1.2. *Cyclodextrins etheric derivatives*

5.1.2.1. *Some important cyclodextrins etheric derivatives*

In application, the most important chemically modified CDs are the ether derivatives, such as hydroxypropyl-, methyl-, ethyl-CDs. These cyclodextrins etheric derivatives (CEDs) with the improved solubility, encapsulation ability and low toxicity have already been commercialized and widely used. The CEDs could be prepared in two ways. One way is that the CD reacts with the free hydroxyls such as dimethyl sulfate, diethyl sulfate, alkyl halides and epoxides. The other way is by sulfonic ester transformation.

5.1.2.2. *Synthesis of the CEDs*

5.1.2.2.1. Hexa-(2, 6-di-O-methyl)-α-CD

α-, β-and γ-CDs 2, 6-bis-O-methyl derivatives are usually prepared by the reaction of dimethyl sulfate and CD. This method should be carried out in alkaline aqueous solution. Another way is by adding the iodine methane to CD DMF solution mixed with barium oxide and barium hydroxide. The raw product contains four components. They could be separated on columns and the pure product yield is 13% [4].

5.1.2.2.2. Heptakis-(2, 3, 6-tri-O-methyl)-β-CD

A synthetic approach is to add the excessive amount of potassium hydroxide and phase transfer catalysts octyl methyl ammonium chloride into β-CD and tetrahydrofuran solution. Then the dimethyl sulfate is dropped in a cooling condition. All the products can be detected by TLC. The yield of dimethyl-β-CD (DM-β-CD) is 60%–70%, and TM-β-CD is 10%–15% [5].

5.1.2.2.3. Octa-(3, 6-bis-O-methyl)-γ-CD

Tert-butyl dimethyl chlorosilane (TBDMS) could migrate between the close hydroxyl of the glucose unit under alkaline conditions. A novel method has been developed to prepare the target product base on the above properties. In doing so, the yield is also improved. The same method could be applied to produce 6-(3,6-di-O-methyl)-α-CD and 7-(3,6-di-O-methyl)-β-CD [6].

5.1.2.2.4. Mono-2-O-(2-hydroxypropyl)-β-CD

Dissolve β-CD in 1.5% NaOH solution. Then, add the mixture of ethyl ester and propylene oxide continuously while stirring in the ice bath. The reaction is kept for 24 h. Thereafter, the reaction buffer is neutralized by HCl. Rest of the unreacted β-CD is precipitated by toluene. Meanwhile, the filtrate is distilled to nearly dry condition. The crude product is extracted by acetone and vacuum-dried to powder. After that, dissolve the product into less DMF and filter the needless inorganic salt. The pure product can be finally precipitated by acetone [7].

5.1.2.2.5. Heptakis-(6-O-tert trimethylsilyl)-β-CD

Dissolve β-CD in pyridine. The pyridine solution which contains tert-butyl dimethyl chlorosilane (TBDMSCl) is dropped into the above β-CD solution in an ice bath. The mixture should be maintained in the ice bath for 3 h before the reaction under room temperature for 12 h. The product should be precipitated by water. The yield is 83.5%. The crude product needs to be recrystallized in methanol [8].

5.1.2.2.6. Total-(2,6-2-O-TBDMS)-γ-CD

Add TBDMSCl and DMAP into the pyridine/DMF solution which contains γ-CD. The reaction is kept under 100°C for 18 h. After cooling, the solvent should be removed by vacuum distillation. The products are distributed in water and dichloromethane layers. Organic layer is washed with potassium hydrogen sulfate and water. The solvent is removed again by vacuum distillation. The yield of target product can reach 85% after column separation [6].

5.1.2.2.7. Multi-(2,3,6-O-4-methyl butyl)-β-CD

The target product can be obtained by the reaction between L, 4-butane sultone and β-CD in alkaline solution. Different reaction conditions can have different average DS. Capillary electrophoresis is always applied to analyze the differences in the composition of product mixture [9].

5.1.2.2.8. Heptakis-(6-O-2-carboxylic acid)-β-CD

Reaction between full (2,3-2-O-acetyl-6-deoxy-6-iodine)-β-CD and 6-hydroxy-2-naphthalene carboxylic acid methyl ester could generate the whole (2,3-di-O-acetyl-6-O-2-naphthyl methyl)-β-CD in the NaH/DMF solution [10].

5.1.3. *CD ester derivatives*

5.1.3.1. *Important CD ester derivatives*

CD ester derivatives are modified by ester bonds. According to the properties of substituent, all the derivatives could be divided into two types, organic and inorganic ester derivatives. Organic ester derivatives are formed by replacing the hydroxyl group by using organic acid chloride class free agent. Some important esters include acid ester, carboxylic acid esters and carbamates. There were few studies on the CD inorganic ester derivatives, including nitrate, sulfate and phosphate.

5.1.3.2. *Synthesis of the CD ester derivatives*

5.1.3.2.1. Mono-(6-O-tosyl)-β-CD

The benzoyl acid ester CD is an important intermediate in CD modification. Selecting the appropriate nucleophile (iodide, azide, thio-amyl acetate and hydroxylamine, alkyl amines) to attack the carbon atoms which connect with tosyl, can trigger nucleophilic substitution reaction. Then, series of functional CD sulfonate derivatives can be obtained.

Dissolve the pre-treated dry β-CD in dry pyridine, and mix with the dry pyridine solution of p-TsCl. Stir rapidly in the ice bath for cooling to 0°C. Keep the reaction solution for 24 h under 4°C, and then for another 24 h at room temperature. Finally, add 1 mL of distilled water to stop the reaction. Most of the pyridine should be removed under a low pressure and below 40°C. The residue is dissolved in hot distilled water to remove the insoluble material after vacuum evaporation. The crude product should be soaked in ether. Filter the liquor to get the white powder, and re-crystallize in water for four times, then vacuum dry for 4 h. The final yield is 32% [11].

5.1.3.2.2. Mono-(2-O-tosyl)-β-CD

Dissolve 8.0 g β-CD in 320 mL NaOH (0.15 mol/L) aqueous solution at room temperature. Drop 8.0 g of the prepared solution in the toluene sulfonyl chloride (p-TsCl) solution which contains little acetonitrile. While stirring the reaction buffer, add 1 mol/L NaOH solution continuously to maintain the pH above 12.5. In order to terminate the reaction, pour 1 mol/L HCl into the buffer. Filter the unreacted p-TsCl and desalt by dialysis. After vacuum evaporation, add 400 mL of methanol. Filter the insoluble components and then vacuum evaporate.

The resulting product needs to be dissolved in hot water, followed by the cooling process, i.e. store in the fridge for 1 week. Filter the white precipitate. Then vacuum evaporate to obtain the final product. The yield is about 53%. After purification the yield is about 42% [12].

5.1.3.2.3. Heptakis-(2-O-tosyl)-β-CD

Dissolve the silylated β-CD in 50 mL pyridine. Add 1.3 g DMAP and 2.1 g p-TsCl and keep the mixture at 50°C for 24 h. Concentrate the reaction solution by vacuum evaporation. Extract the residue by ethyl acetate. After washing the filter cake by HCl and NaHCO$_3$ continuously, the product is purified by column separation and the yield is about 58% [13].

5.1.3.2.4. Mono-(3-O-tosyl)-α-CD

Dissolve 4 g α-cyclodextrin in 5 mL water. Adjust the pH value to 12 by NaOH solution. And mix with 1.4 g p-TsCl. Vigorously stir the mixture at room temperature for 5 min. The pH value would decline to 6.5. Filter the reaction solution, and separate the products by reversed-phase column. Three kinds of purified products would be detected. Mass spectrometry and NMR data showed that the three kinds of products were C-3, C-2 and C-6 position substituted toluene sulfonyl α-CDs [14].

5.1.3.2.5. Mono-(3-O-β-naphthyl sulfonyl)-β-CD

Add 3.5 g β-naphthalene chloride into 50 mL 30% acetonitrile aqueous solution (pH = 12) which contains 3.5 g β-CD. Vigorously stir the reaction buffer till the pH value reaches 7.0. Then filter the solution by reverse phase column. The product yield is about 20% [15].

5.1.3.2.6. Mono-(6-O-benzoyl)-β-CD

Dissolve 25 g of dried β-CD in 750 mL evaporated pyridine in ice bath. Then add the prepared solution into 70 mL benzoyl chloride solution which contains 2.4 g and dried for 20 min. Stir the reaction mixture at room temperature for 5 h. After that, stop the reaction by adding a little water. Then steam at 40°–50°C to remove pyridine. Extract the residue successively with acetone and methanol and wait for them to precipitate as white solid powder at room temperature. The white crude product should be recrystallized in methanol/water to get pure samples. The final yield is 17% [11].

5.1.3.2.7. Mono-(2-O-benzoyl)-β-CD

Add 100 mL acetonitrile solution which contains 1 g benzoyl chloride to 350 mL β-CD (6.5 g) solution (adjusted to alkaline with NaOH). The reaction lasts 5 min and the pH would decrease to 7.0. Evaporate the solvent, and then the residue is precipitated by 500 mL acetone to obtain crude product. The product can be purified by silica gel column. The yield is 8% [16].

5.1.4. *CD modified by special functional groups*

5.1.4.1. *CD derivatives with embedded functional groups*

CD derivatives have wide applications in analogue enzyme and analysis chemistry. CD derivative with special functional group can be used as a simple enzyme model which is convenient for studying the action mode and functional groups of the enzyme. The fluorescent and chromophore groups linked to the CDs can be seen as an indicator or a probe which are beneficial for separating the isomers and enantiomers.

5.1.4.1.1. Introducing the small biomolecules

Peptides, coenzymes, amino acids are special functional biological molecules. They became a hot topic while studying the analogue enzymes. For example, the ALT model was formed by connecting the cofactor pyridoxamine phosphate (pyridoxamine, vitamin B_6 amine) and ethylenediamine to β-CD. The two functional groups were located through the intermediate positioning process.

Tripeptide Ser–His–Asp had been introduced to the CD for simulating the chymotrypsin by the following method. Mix 2-amino ethyl mercaptan salts, 6-O-tosyl-β-CD and NH_4HCO_3 with DMF/H_2O (1:3) solution under nitrogen protection at 60°C for 4 h. After vacuum concentration, separate the product by Sephadex G-10 column. The yield is about 43%. Mix the above product with Boc-Ser (Ot-Bu) -His–Asp (Ot-Bu) OH, dicyclohexyl carbodiimide, 1-hydroxy triazole in DMF solution to obtain 6-[Boc-Ser (Ot-Bu) -His-Asp (Ot-Bu) NHCH$_2$CH$_2$S]-β-CD. Then treat with the TFA/CH$_2$Cl$_2$ (9:1) mixed solution, vacuum concentrate and freeze-dry. This modified CD can be used as an analogue enzyme to hydrolyze nitrophenol esters.

In addition, some researchers have tried to introduce the enzyme nicotinamide drug molecules into the CD skeleton to deliver the drug to specific locations.

5.1.4.1.2. Introducing the chromophore

Aromatic, dye and other chromophore-containing molecules have a variety of special physical properties. CDs modified by these groups exhibit many new properties. Since the chromophore can be used as a spectral probe, this makes it possible to apply UV, fluorescence, circular dichroism spectroscopy and other means in the molecular recognition field. For example, CD phosphate has a positive electron deficient center, which is a new recognition site for multiple identification of amino acids.

5.1.4.1.3. Introducing the fluorophores

Since the fluorophores-modified CDs can be used as the probe and fluorescent sensors for detecting various guest molecules, they have been researched a lot.

Keep N, N′-dicyclohexyl carbodiimide and 4-(1-pyrene)-butyric acid in DMF solution at −4°C. Add 6-deoxy-6-amino-γ-CD to the above reaction buffer at room temperature. Then pour the mixture into acetone and recover precipitate. The mono-[6-(1-pyrene)-Ding amido-6-deoxy]-γ-CD can be obtained. Finally, the crude products can be purified through eluting by DMF/H$_2$O in the Sephadex LH-20 column.

The fluorescent CD, N-fluorescein-N'-(mono-6-deoxy-6-β-CD) thiourea, could append to the water-soluble polymer chain to form light-emitting "light" elements collar, or fluorescence quasi-linear alkyl.

5.1.4.1.4. Capped CDs

At first, the substituent replaced one end of the CD hydroxyl. Then, through multi-point connection of hydroxyl sites or stereochemical structure of the substituent itself, capped CDs, i.e. CDs with a cover of the substituent, could be formed. Capped CDs can be used to enhance the inclusion ability of the CD and improve molecular recognition of the specificity guest. A typical example is the basket CD which can be obtained by the reaction between biphenyl-4, 4'-di chloride and β-CD at A, D-position, resulting in sulfonic esters. Then the reaction with potassium iodide to form the di-iodo-β-cyclodextrin. di-iodo-β-CD can be precipitated by using H$_2$O / (CH$_2$CH$_2$)Cl$_4$ (20:1) at 0°C. After reacting with 10-fold excess of aza-crown ether in dry DMF under the protection of the chlorine reaction at 55°C for 24 h, the solvents are removed by evaporation. Purification can be achieved by sephadex LH-20 column with DMF/H$_2$O (1:1).

5.1.4.2. *CD dimer (bridged-CD)*

CD dimer is also known as bridged CD (bridged β-CDs). There are problems for CDs to simulate the specific binding of an antibody and the specificity and catalytic activity of an enzyme. One is the low binding constant. The other one is the large flexibility when used as an ideal enzyme stimulate. The binding ability and molecular recognition of the bridged CDs are better than that of the related native CDs. The dimmer could increase the encapsulation and molecular recognition ability as binding sites increased and structure changed. It is important to investigate the analogue enzyme and the formation of order structure. The artificial analogue enzyme, molecular recognition, drug delivery, chromatography have been developed already.

CD dimmers includes different types: (1) Flexible alkyl chains; (2) Rigid aryl, such as benzene, naphthalene, biphenyl, imidazole, thioxanthone and so on; (3) Disulfide bonds; (4) Chiral centers; (5) Metal ligand.

The bridge pathways could be divided into single and double bridge abutments. There are many conformations while the two CD molecules are connected by single bridge. The double bridge abutment CDs refer to two neighbor CD molecules that are linked by two bridges and usually have two kinds of isomers.

5.1.5. *CD polymer*

CD polymers (CDPs) are macromolecule derivatives which carry multiple CD units. These units are appended by chemical bonding or physical mixed method. These kinds of polymers not only have the better ability of identification and encapsulation, but also possess good mechanical strength, stability and chemical tenability. The comprehensive function of CD cavity and polymer network could improve the properties of materials. The early investigations always focused on the formation of CDPs by monomers copolymerization. However, several papers showed that the CDs could link to the natural macromolecules to form novel polymers. These polymers were expected to be applied in the functional materials, separation and analysis technology, biomedical engineering, environmental and other high-tech fields [17].

5.1.5.1. *Synthesis pathways of CDPs*

5.1.5.1.1. Crosslinking

Crosslinking is referred to the copolymerization between the hydroxyl of CD and the di-functional groups or multi-functional groups of a compound for example: epoxide. Crosslinking can be divided into bulk polymerization and copolymerization. Bulk polymerization occurs when the monomers are cross-linked directly by the crosslinking agent. Epichlorohydrin (EPI) is the most common crosslinking agent for bulk polymerization.

In addition to using the epoxy compound, ester reaction also could be used to synthesize CDPs by crosslinking. For example, β-CD could be copolymerized by reacting with the polyacrylic acid. CDs are modified by chemical modifications to generate new CD derivatives for further crosslinking. The copolymerization of these derivatives are then triggered under appropriate conditions. Crosslinked polymers always contain high content of CD. Preparation of the polymers is simple, but the mechanical strength and chemical stability is poor. Especially in the chromatographic separation, the low efficiency restricts the application of the above mentioned polymers.

5.1.5.1.2. Immobilization

Immobilization is referred to the CDs that are linked to the macromolecule skeleton by chemical bonds. The basic skeleton could be polystyrene, polysiloxane, poly-ethylene, cellulose, chitosan and so on. According to the different synthetic routes, the immobilization methods can be divided into the following types. (a) Before immobilization of the CDs to the skeleton, a series of special functional groups are attached to the skeleton. Then, the related CDs react with the group to form the target polymers. (b) First, modify the CDs to monomer CD derivatives by the chemical method, and then immobilize the monomer derivatives to the skeleton. (c) The macromolecular carrier and the CDs would need to be modified before immobilization.

The widely used carrier is silica gel. Researchers tend to modify the silica gel or CD monomer first to make an arm for convenient immobilization. Immobilized CD polymers are stable, firm and the chemical structure could be adjusted. However, because of the large steric hindrance of CD molecules, the immobilization yield is low.

5.1.5.1.3. Combined action of crosslinking and immobilization

Because crosslinking and immobilization both have some drawbacks, a new method that combines immobilization and crosslinking as been developed. The novel CDPs exhibit good mechanical strength and high immobilized ratio. But the synthesis process is complicated.

5.1.5.1.4. Blend

The technique of blending refers to dissolving the CDs and the macromolecular polymers in a solvent and then evaporating the solvent, or by mechanical blending to get the CDPs. No chemical reaction is involved. Strictly speaking, the resulting product is not the real CDP. However, because of the simple operation and wide application, the technique of blending has received a lot of attention.

5.1.5.2. *Application of CDP*

5.1.5.2.1. Industrial application

Since some essential components of the cosmetics and condiments are unstable and volatile, CD-modified super water-absorbing starch resin was applied to absorb these components. For a long time, CDPs have been used as the encapsulation material for including food additives, cosmetics and condiments.

In 1996, Wacker–Chemie–GmbH Co. Ltd in Germany developed an active CD, and immobilized it to the cellulose by dipping and padding. This kind of novel CD could be applied to modify not only the cellulose, starch, gelatin and other natural polymer materials, but also the polyester, polyamide, polypropylene and other synthetic fibers. Therefore, they could be used in different functional materials if treated appropriately. For example, a kind of fiber-CDPs encapsulated squalane was developed in Japan. Despite repeated washing, encapsulated squalane was able to maintain the moisturizing capabilities.

The CD fiber polymers also could be used as an essence storehouse while applying to the textile material. The flavor molecules are encapsulated in the polymers for improving the stability. The smell can be retained for months or longer.

5.1.5.2.2. Environmental protection

Since the CDPs could capture the organic molecules and complex the metal ions from the liquids, they are used in the field of environmental monitoring and wastewater treatment. If the pesticide is encapsulated in the insoluble polymers, it will not be washed out by rain.

5.1.5.2.3. Medicine

Polymer drug delivery systems (DDS), such as microspheres, nanospheres, microcapsules, etc, are developed to improve the medicine's release ability. CDs could complex with many drugs via the noncovalent bond. However, the CD itself had some shortcomings, thus the native CDs should be modified. While the CDs are added into the DDS by physical mix and noncovalent binding, the delivery mechanism of DDS can be improved.

5.1.5.2.4. Separation and analysis technology

Cavity size of the CD molecules may match with the guest molecules. The hydroxyl located in the cavity edge can bind guest molecules via hydrogen bonds. And the hydrophobic cavity will trap the fat-soluble molecules. Therefore, it is successfully applied to a variety of chromatographic and electrophoretic methods for separating isomers and enantiomers coursed by the identification and selection ability. Many drugs or their precursors contain at least one chiral center, and commonly, one enantiomer is active while the other one is not even toxic. Using CDPs will greatly improve the enantiomer separation efficiency. Especially, the CDs which are modified by chiral molecules could be used to recognize the enantiomers.

5.2. Preparation of Methylated CD

5.2.1. *Methylated CD*

Methyl-β-CD is generated by methylation of β-CD. This reaction only occurs in alkaline condition. The methylation is at 2, 6-position due to the steric crowding. However, the 3-position also can be replaced under extreme condition. The main product of methylation reaction is hepta-(2, 6-di-O-methyl)-β-CD. The ratio is up to 50%. The solubility of methylated CDs is considerably higher than that of the natives. About 55 g dimethyl-β-CD can dissolve into 100 mL water to reach saturation. The trimethyl-β-CD is 30 g in 100 mL water. The solubility of dimethyl-β-CD is 15 times more than β-CD in ethanol. With the substituent number increasing, the solubility of MCD rises till the substituent number reaches to 13 or 14. The solubility of native CD always increases with increase in temperature. On the contrary, less MCD dissolves in water under higher temperature. MCD also has some properties, such as low moisture absorption and high surface activity. It can be used to include the guest with a better solubility of complexes. The methylation not only improves the physical properties of CDs, but also changes the structure of CDs. X-ray diffraction results show that the structure of trimethyl-β-CD is different from the native β-CD. The original narrow opening hole became narrower, and wide openings became wider. MCD also has stereo-selective property due to its increased steric hindrance [18].

5.2.2. *Chemical preparation method of MCD*

The most common way to produce MCD is the dimethyl sulfate method. The mechanism of reaction is as follows:

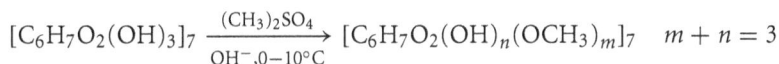

$$[C_6H_7O_2(OH)_3]_7 \xrightarrow[OH^-,0-10°C]{(CH_3)_2SO_4} [C_6H_7O_2(OH)_n(OCH_3)_m]_7 \quad m+n=3$$

Dissolve 2 g sodium hydroxide into 30 mL deionized water which contained 10 g β-CD. Then add 5 mL dimethyl sulfate slowly with stirring. Keep reaction system at 60°C for 8 h. After that, cool the reaction buffer and adjust to neutral using diluted HCl. Then remove anions and cations through ion exchange resins (repeat 2 or 3 times). Concentrate the collected liquid to a syrup-like state by vacuum distillation. Finally, disperse the product in 40 mL acetone and vacuum dry to get the product.

At present, the preparation of fully replaced MCD is mostly one-step. The detailed process is as follows: Load 30 mL dimethyl sulfoxide (DMSO) in 50 mL-three neck flask. Add β-CD and stir for 1 h. After β-CD is fully dissolved, add 0.6 g sodium hydride under the protection by nitrogen. Drop double amount of methyl iodide to the buffer. The reaction lasts for 5 h at 60°C. Stop the reaction and add sodium hydride. Extract the product by ether three times. Evaporate the ether, get the crude product, wash it using chloroform and methanol (100:2) and separate via silica gel column. The final yield could reach to 54%.

5.2.3. *The separation and purification of MCD*

The industrial MCD is a mixture of different kinds of partly replaced methyl CD derivatives. The separation and purification process is relatively simple, and mainly aims at removing the salts and unreacted raw materials from the reaction solution. Salt is removed by ion exchange resin for several times. The un-reacted raw materials are usually neutralized by ammonia water or steamed [19].

For further investigation, the product needs to be purified via silica gel column with the buffer of chloroform/methanol (100:2).

5.2.4. *Analytical techniques for MCDs*

5.2.4.1. *Fourier transform infrared (FT-IR)*

Mix a small amount of methyl-CD products with KBr. Grind the solid mixture and tablet for detection. The difference in structure and formation could be confirmed by comparing with the native CDs. The hydroxyl absorption of methylated CDs is obviously weakened at 3,200–3,700 cm^{-1} since the methyl broke the hydrogen bond of CD molecule. The characteristic peak of methyl at 2,934, 1,461 and 1,368 cm^{-1} can demonstrate that the methylated CDs formed successfully.

5.2.4.2. *TLC*

The MCD mixture could be determined and purified by thin-layer chromatography. Dissolve methyl CD and native CD and then spot on the silica gel chromatography plate. The developing agent is 100:2 chloroform/methanol mixture solvent. A higher DS resulted in the longer developing distance.

5.2.4.3. *Thermal analysis*

Solid-state CD and its derivatives could be analyzed using the TG method. Weigh 20 mg sample in the copple, and heat slowly (10°C/min) under protection of inert gas. The scan scope ranged from 0°C to 800°C. The thermal degradation kinetics, degradation temperature and thermal stability could be investigated through the analysis of TGA/DTA spectrum.

5.3. Preparation and Analysis of Hydroxypropyl-CDs

5.3.1. *Hydroxypropyl-CDs*

Among the three major native CDs, the internal cavity of α-CD is the smallest. Usually, it can only encapsulate smaller guest molecules, and its applied range is small. γ-CD has a big cavity which can encapsulate bigger guest molecules, however, the application of γ-CD is limited due to its high cost production. β-CD is the most accessible, the lowest-priced and generally the most useful, but β-CD is not very soluble in water due to the strong hydrogen bond between HO-2 and O-3 [20].

Hydroxylpropyl-β-CDs (HP-β-CDs), a hydroxyalkyl derivative, is an alternate to native CD, with improved water solubility and may be slightly more toxicologically benign. As the first approved CD derivatives by FDA, HP-β-CDs have wide applications in food, agriculture and the pharmaceutical fields.

HP-β-CDs are prepared by reacting β-CD with propylene oxide in alkaline aqueous solutions. The reaction mechanism is dimolecular nucleophilic substitution reaction (SN2). The CD alkoxyanion attacks the carbon atom which has the least substitutions in the oxygen ring, because the steric hindrance is minimum there. Therefore, propylene oxide ring open gaining the 2-hydroxylpropyl substitutes. The high alkali concentration favors alkylation at O-6, while the low alkali concentration favors alkylation at O-2. The products are always substituted randomly when it comes to distribute among the different glucose units, and the reaction products are always amorphous compounds [21]. There are three principal structural types of HP-β-CDs. The substituents may be distributed equally between the glucoses of CD, clustered on one glucose, or form an oligomeric chain growing from the CD. The first type was considered the most probable since the reactivity of corresponding primary hydroxyls in every glucose residue is equal. If all the hydroxyls in HP-β-CDs have the same reactivity, the probability of further substitution on an already substituted glucose residue is only one in seven.

For many years, CDs were considered to have rigid truncated-cone structures although this concept was incompatible with the ease of their complex formation and several experimental findings (mainly obtained by NMR technique). Recent experimental results demonstrate that the complexes, held together by weak nonbonded interactions, must be flexible not only in solution but also in the solid state. The truncated-cone structure of HP-β-CDs became more flexible since the intramolecular hydrogen bond is broken by the hydroxylpropyl substituent. This flexibility preferably explains the improvement of complex forming ability and dynamic character of the complexes of HP-β-CDs [22].

The toxicity of HP-β-CDs is lower than parent β-CD. There have been a large number of toxicological and pharmaceutical technological experiments and numerous human clinical trials which demonstrated that HP-β-CDs can be used in oral medicines and injections. This CD derivative has been approved by FDA. HP-β-CDs is well tolerated in the animal species tested (rats, mice and dogs), particularly when dosed orally, and shows only limited toxicity. In short duration studies, there are slight biochemical changes, whereas in studies of a longer duration, up to three months, additional minor hematological changes occur but with no histopathological change. When dosed intravenously, histopathological changes were seen in the lungs, liver and kidney but all findings were reversible and no effect levels were achieved. The carcinogenicity studies showed an increase in tumors in rats in the pancreas and intestines which were both considered to be rat-specific. There were also non-carcinogenic changes noted in the urinary tract, but these changes were also reversible and did not impair renal function. There were no effects on embryo–fetal development in either rats or rabbits. Humans have good tolerance for HP-β-CDs. The main adverse effect of HP-β-CDs on the human body is diarrhea and there is no adverse effect on kidney function [23].

The water solubility of HP-β-CDs is greatly increased (>500 g/L, $20°C$) because of substitution, and is hygroscopic. Usually, HP-β-CD is a white amorphous powder. However, if the DS of HP-β-CDs is above 10, it is hard to dry them. In this case, HP-β-CDs are like transparent sticky gels. The DS of HP-β-CDs is a mean value. The chemical property of HP-β-CDs is similar to native β-CDs and the free hydroxyl groups can also react with other substitution reagents. DS and substitution pattern affect the complex forming ability of HP-β-CDs.

5.3.2. *Preparation of hydroxypropyl-CDs*

The common method used in the preparation of HP-β-CDs is that the β-CD reacts with propylene oxide in alkaline aqueous solutions, and the product is obtained by concentrating and drying [24]. A detailed process is as follows: heat 75 mL sodium hydroxide (18%, m/m) to 60°C; add β-CD (50 g) to the solution with agitation until β-CD dissolves completely. Cool down the solution to room temperature; add 25 mL propylene oxide evenly in 3 h in freezing point with agitation. Then, heat the mixture to a designated temperature, keep on reacting. At the end of the reaction, add HCl to stop the reaction (pH=7). Then, distill the mixture under reduced pressure to gain a thick slurry. Dissolve the slurry in alcohol (95%, v/v), intensely agitate, precipitate with NaCl on standing. Filter the mixture, and distil the liquid under reduced pressure. Dialyze the resulting thick slurry in water, then, distill the solution and freeze dry to gain the HP-β-CD samples.

Chao Yuan [25] has optimized the preparation process of HP-β-CDs. Reaction time (A) reaction temperature (B), dialysis time (C) were employed in the single factor experiment. Then, a central composite design of response surface methodology was used with the DS and yield response values.

According to the results of the single factor experiment, reaction time was set in the range of 10–20 h, reaction temperature 20°C–40°C, dialysis time 6–10 h, DS and yield were employed as response value. A response surface method experiment was designed by a software — Design Expert 6.0. The results are shown in Table 5.2.

The designed experiment was carried out and the products were obtained. The response values of the products were determined. Then, the results were analyzed, a quadratic polynomial was chosen as the simulation equation and the optimized dynamic parameter equation of response values are shown in Table 5.3.

The result of variance analysis showed that the simulation equations of DS and yield were both significant. For DS, the F value of the regression model was 5.79, $P < 0.05$, the model was significant, $R^2 = 0.9125$, indicated a fine accuracy of the model, P value of lack of fit was 0.0024, fitting effect was good, this model should be used for forecasting. For product yield, the F value of the regression model was 24.77, $P < 0.01$, the model was highly significant, $R^2 = 0.9781$, indicated a fine accuracy of the model, P value of lack of fit was 0.2839, fitting effect was not so good, but this model also could be used for forecasting.

Table 5.2.   Response surface methodology design and results [25].

Run	A: Reaction time (h)	B: Reaction temperature (°C)	C: Dialysis time (h)	DS	Yield (%)
1	10.00	20.00	6.00	3.42	56.40
2	15.00	30.00	8.00	3.96	63.20
3	22.07	30.00	8.00	3.86	53.10
4	7.93	30.00	8.00	3.65	49.60
5	20.00	20.00	10.00	3.22	38.60
6	20.00	40.00	6.00	3.85	50.10
7	15.00	30.00	10.83	3.77	49.00
8	15.00	30.00	8.00	4.02	62.70
9	15.00	30.00	5.17	3.85	60.10
10	15.00	44.14	8.00	3.78	47.50
11	10.00	40.00	10.00	3.12	35.10
12	15.00	30.00	8.00	4.00	60.40
13	15.00	30.00	8.00	3.91	59.20
14	15.00	30.00	8.00	3.95	58.90
15	15.00	15.86	8.00	3.61	50.70

Table 5.3.   Response surface model [25].

Response value	Two polynomial regression model	R^2	Pr> F
DS	$DS = -1.0131 + 0.1138A + 0.0878B + 0.6754C$ $-0.0068A^2 - 0.0020B^2 - 0.0355C^2 + 0.0041AB$ $-0.0022AC - 0.0029BC$	0.9125	0.0337
yield	$yield = -10.1426 + 2.9706A + 1.6759B + 7.8981C$ $-0.2114A^2 - 0.0641B^2 - 0.9215C^2 + 0.0855AB$ $+0.1317AC + 0.0969BC$	0.9781	0.00135

The response surface plot is a three-dimensional space surface which is formed by the response value of interaction of test factors. The effects of test factors on the response value can be found by analyzing the response surface. The effect of interaction on DS is shown in Fig. 5.2. In the plot, the interaction between reaction time and reaction temperature had a major effect on the DS for the quick drop of the response surface and the serried contour line. Dialysis time had less effect on the DS. The design point was gained under the reaction time 15 h, reaction temperature 30°C, dialysis time 8 h and the DS was 4.0. Figure 5.3 shows the effects of interaction on yield of HP-β-CDs. The contour line of dialysis

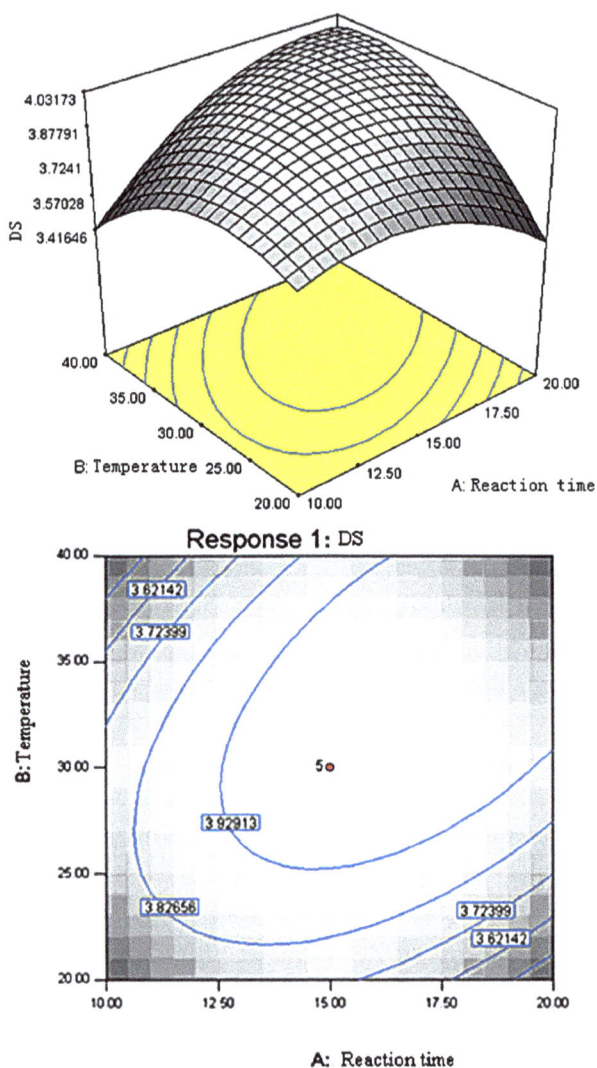

Fig. 5.2.   Effect of reaction time — temperature interaction on DS [25].

time was the most serried, thus, dialysis time was the dominating factor for the yield. The yield of HP-β-CDs declined sharply along with the extension of the dialysis time. Therefore, dialysis time should be controlled in 8 h. From the response surface plot, it was found that all the best working points of the interaction fall in the range of the designed test. This demonstrated that the response surface

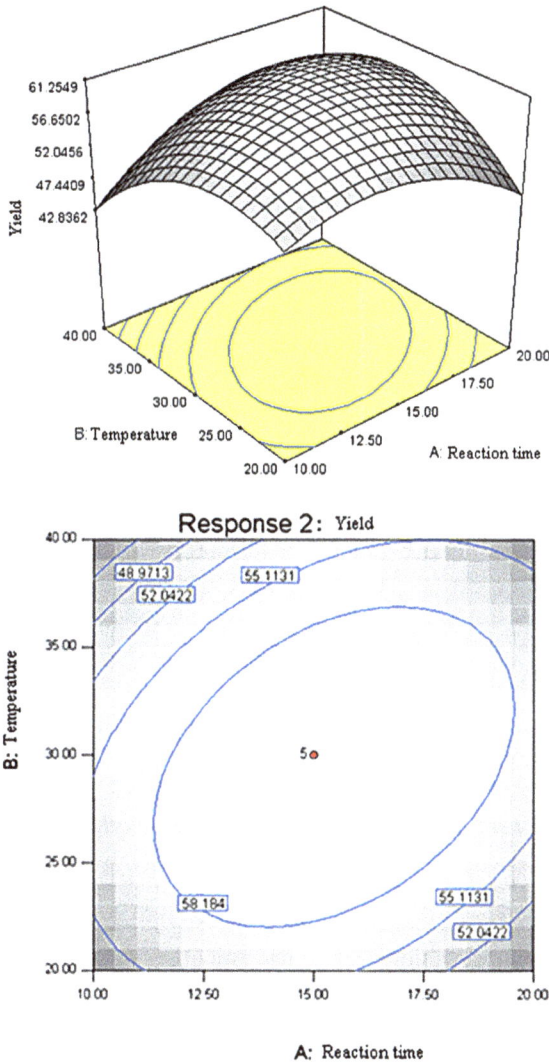

Fig. 5.3. Effect of reaction time — temperature interaction on yield [25].

experimental design was reasonable. Considering both the requirements of the DS and yield, the optimized experimental parameters were: reaction time 16.05 h, reaction temperature 31.06°C, dialysis time 7.41 h and the DS was 4.0, yield 61.8%. Through tests and verifications, under above optimized conditions, the real DS and yield were 4.1 and 60.23%, respectively. The results agreed with the theoretical conclusion. It should be pointed out that dialysis was used in this experiment for obtaining pure products. However, in industrial production, ion exchange may be more suitable to purify the products.

5.3.3. *Separation and purification of hydroxypropyl-CDs*

HP-β-CDs amorphous mixture is more suitable in inclusion application. The amorphous product has a better dispersion property and solubility, and all the HP-β-CD series is non-toxic, thus the preparation conditions and purification process of HP-β-CDs could be simplified; only the salts and the raw material which are not reacted should be removed from the reaction liquid. Salts generally are removed by the ion exchange. The HP-β-CD products are obtained through the acetone precipitation method.

Further purification of HP-β-CDs can be carried out through the thin layer chromatography. A formula for developing was: propanol, ethyl acetate, aqua ammonia and water were prepared by 6:1:1:3, or acetonitrile, water, aqua ammonia was mixed by 6:3:1, the latter formula has a shorter developing time. The color development reagents are: cerous sulfate 1 g, ammonium molybdate 24 g, concentrated sulfuric acid 50 mL, 450 mL deionized water. Because of the adsorption principle of the silica gel, the more the substituent on the HP-β-CDs, the faster the molecules will move. Therefore, the high DS HP-β-CDs move a long distance. Furthermore, HP-β-CDs can be purified by silica column. The mobile phase was chloroform/methanol (100:1) [26].

5.3.4. *Analysis and detection technology of hydroxylpropyl-CDs*

5.3.4.1. *Infrared spectroscopy*

Infrared spectra of β-CD and three HP-β-CDs with different DS are shown in Fig. 5.4. All the samples exhibit the characteristic absorption peaks of carbohydrate: 3,400 cm^{-1} (O–H stretching vibration), 2,930 cm^{-1} (C–H stretching vibration) and 1640 cm^{-1} (O–H the bending vibration), 1,155 cm^{-1} (C–O vibration), etc. There is a characteristic absorption peak of α type glycosidic bond in about 855 cm^{-1} which proves that the CDs are formed by α-1, 4-glycosidic bond. CD's skeleton structure has not changed after modification. The strong absorption peak at wave number 3,400 cm^{-1} belongs to hydroxyl. This peak is wide and strong due to the formation of intramolecular hydrogen bond which decreases the force constant. The peak of β-CD is wider than those of HP-β-CDs which means that hydroxypropyl opened the hydrogen bonds in β-CD. The 2,960 cm^{-1} peak is an antisymmetric vibration peak of methyl. β-CD had no peak at this place for it did not contain methyl. HP-β-CDs have a methyl in the hydroxylpropyl, and a

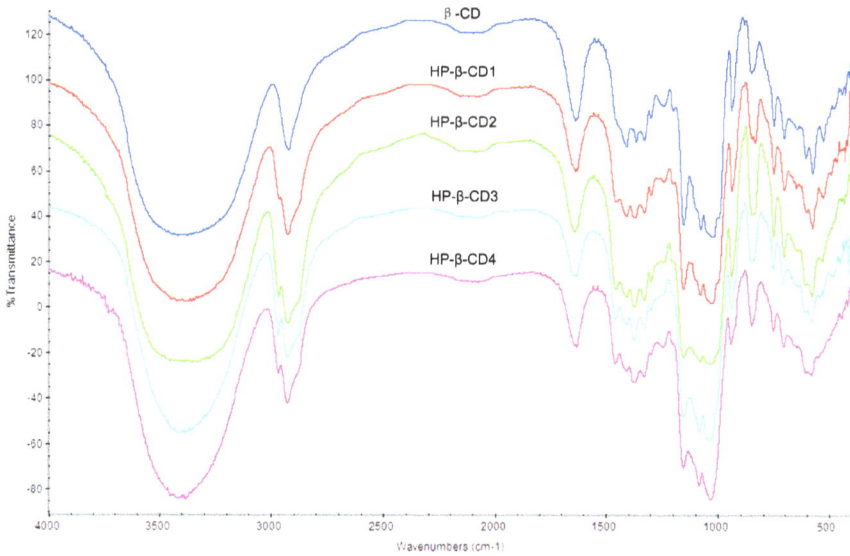

Fig. 5.4. Infrared spectra of β-CD and HP-β-CDs [32].

peak at 2,960 cm^{-1}. This proves the exit of hydroxylpropyl. The area of this peak is increased from HPCD 1 to HPCD 3. It means that the DS of the HP-β-CD is increased. In addition, the infrared spectrum of HP-β-CD has a symmetric bending vibration peak of methyl at 1,375 cm^{-1} which is the characteristic peak of methyl. Infrared spectra of the three HP-β-CDs have the same characteristic peaks with the standard, indicating that they are the same substance, and the three HP-β-CDs contain no impurity [27].

5.3.4.2. *Thin layer chromatography*

As described in Sec. 5.3.3, thin layer chromatography could be used in the purification of HP-β-CDs. Nevertheless, thin layer chromatography could be used in qualitative analysis of hydroxypropyl derivatives of CDs including composition and DS. Thin layer board is a silica gel plate which can be purchased or prepared in lab. The formula for developing are: propanol, ethyl acetate, aqua ammonia and water are mixed at a ratio of 6:1:1:3, or acetonitrile, water, aqua ammonia are mixed at a ratio of (6:3:1). The latter formula has a shorter developing time. The color development reagent: cerous sulfate 1 g, ammonium molybdate 24 g, concentrated sulfuric acid 50 mL are dissolved in 450 mL deionized water.

5.3.4.3. *Thermal analysis*

Modern thermal analysis technology has been applied in many fields. It can be used for measuring the crystalline transition, melting, burning and dehydration, thermal stability, sample purity and dynamic parameters, etc. Most common thermal analysis technologies are differential thermal analysis (DTA), differential scanning calorimetry (DSC), thermogravimetry and thermal mechanical analysis. Since the 1930s, thermal analysis technology has been used in chemistry, especially in assay of thermal stability, purity and dynamic parameters of chemical materials, drugs and food composition. In recent years, the application scope of thermal analysis has almost involved the entire chemistry field and other related disciplines. HP-β-CD is a new type of chemical material with a special structure and function, which has important applications in pharmacy, food, commodity and analytical chemistry. Therefore, the studies on thermal stability and thermal degradation dynamics of HP-β-CDs are of great significance. At present, thermogravimetric analysis (TGA) and DSC are hot methods in the accurate study of thermal degradation of solid CD and its complex. The thermogravimetric/differential thermal analysis (TGA/DTA) is the most appropriate. TGA/DTA can test the change in weight and thermal the in the temperature rising process at the same time. TGA/DTA can be used in testing thermal stability of HP-β-CDs, measure the dehydration and melting process of samples by thermogravimetric curve and differential thermal curve and cross-checking [28, 29].

The TGA and DTA curves of HP-β-CDs are shown in Fig. 5.5. HP-β-CDs exhibit no weight loss before 300°C, indicating that the moisture content of HP-β-CDs is low with no volatile impurities. There is an obvious weight loss between 300°C–400°C. It is caused by the decomposition of HP-β-CDs. Carbon decomposition products were volatilized, which resulted in weight loss. The sample

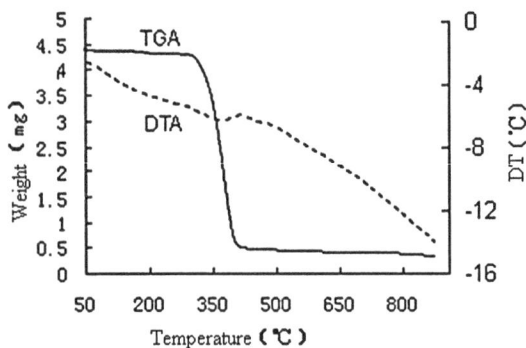

Fig. 5.5.   TGA and DTA curves of HP-β-CDs [27].

loses weight very slowly after 400°C, and HP-β-CDs degrad completely. Only a heat resisting carbon core resisted. From the DTA curve, an endothermic peak can be found between 300°C–400°C with a peak value of 367°C. No other peak can be found on the curve. It means no other physical or chemical process, which can cause the heat change, occurred. The fusion and degradation of HP-β-CDs can be completed in one step [27].

5.3.4.4. *DS and distribution of substituents detection*

HP-β-CD is an amorphous mixture of CD hydroxylpropyl derivant, which contains different components with different DS and distribution of substituents. It is important to analyze the DS and distribution of substituents for understanding the performance of HP-β-CDs. There are several methods in the determination of DS of HP-β-CDs. Chemical method is simple, does not need special equipment. Mass spectrometry and NMR method are rapid and accurate. These three major methods will be introduced in the following text.

5.3.4.4.1. Chemical method

The principle of chemical method to determine the DS of HP-β-CDs is:

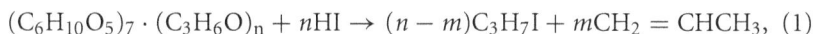

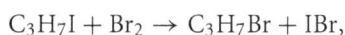

$$(C_6H_{10}O_5)_7 \cdot (C_3H_6O)_n + nHI \rightarrow (n-m)C_3H_7I + mCH_2 = CHCH_3, \quad (1)$$
$$C_3H_7I + Br_2 \rightarrow C_3H_7Br + IBr,$$
$$IBr + 2Br_2 + 3H_2O \rightarrow HIO_3 + 5HBr,$$
$$HIO_3 + 5HI \rightarrow 3I_2 + 3H_2O, \quad (2)$$
$$CH_2 = CHCH_3 + Br_2 \rightarrow BrCH_2CHBrCH_3. \quad (3)$$

Equation (1) shows that the sum of propylene and iodopropane which generated in the reaction between HP-β-CDs and hydroiodic acid is equal to the hydroxylpropyl number of HP-β-CDs. Equations (2) and (3) indicate that one iodopropane molecule consumes three bromine molecules and creates three iodine molecules. One propylene molecule consumes one bromine molecule.

Test equipment is showed in Fig. 5.6.

First, 10 mL sulfuric acid solution (10%, w/w) and 0.5 g red phosphorus were added into a purifying tube. Then, 10 mL potassium iodide (20%, w/w) was put in a tailtube. Finally, the HP-β-CDs sample, 5–6 mL hydroiodic acid and a few grains of zeolite were put in sample flask. All the lines were connected. CO_2 was fed through the side mouth of the flask. The airflow was controlled in 50–70 bubbles per minute. The reaction was kept at 160°C for about 2 h. After the reaction, the bromine solution which was placed in the absorbing reaction tube was discharged

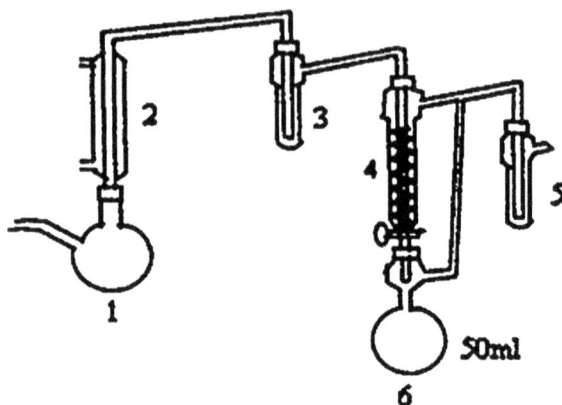

Fig. 5.6. Test equipment [30].

1: sample flask, 2: condenser, 3: purifying tube, 4: absorbing reaction tube, 5: trail tube, 6: accepting flask.

into the connected accepting flask, and cleared with distilled water. The total liquid volume in the accepting flask was kept at about 50 mL. The accepting flask was plugged and shaken. 25 mL bromine solution was taken out by a pipette and placed into an iodine flask for the determination of iodine propane. 10 mL potassium iodide was added into rest of the 25 mL solution in the accepting flask, was shaken, then moved into another iodine flask. 5 mL sulfuric acid (10%, w/w) was added. The solution was titrated by sodiumthiosulfate standard solution employing one or two drops of starch solution as indicator. The volume of sodiumthiosulfate standard solution consumed in the titration was marked as a. In the same way, the potassium iodide in the tailtube was titrated and the volume of sodium thiosulfate standard solution consumed in the titration was marked as b. 10 mL sodium acetate solution (25%) was added into the 25 mL bromine solution which was taken out beforehand, shaken and formic acid was dropwise added until the residual bromine ran clear. A few drops of formic acid were added again. 1 g potassium iodide was added, shaken, then kept for about 10 min, 10 mL sulfuric acid (10%, w/w) was added, titrated and the volume of sodium thiosulfate standard solution consumed in the titration was marked as c.

Then, the test was done again according to above steps without adding the HP-β-CD sample. The quantity of bromine lost in the reaction was marked as $W1$, the blank in iodine propane part was marked as $W2$. In addition, the blank of bromine solution itself, W should be calibrated at any time during the determination.

During the titration, if potassium iodide was added into the bromine solution first, the difference between its titre value and the titre value of blank bromine solution was the quantity of bromine which was consumed by propylene. This was because the quantity of bromine consumed by propylene was counteracted by the quantity of iodine consumed by propylene. When the remaining bromine in the bromine solution was removed by formic acid first, followed by the addition of potassium iodide, the titration result was the quantity of iodopropane.

The value of hydroxypropyl which is determined as propene is:

$$u = 29 \times M[W - (a \times 2 + b + W1)]/(10 \times m). \qquad (4)$$

The value of hydroxypropyl which was determined as iodopropane was:

$$v = 9.67 \times M(c \times 2 - W2)/(10 \times m) \qquad (5)$$

The hydroxypropyl DS of the product was:

$$n = (u + v)/[100 - (u + v)] \times 162/58 \times 3 \qquad (6)$$

In the three formulas above, M is the real concentration of sodium thiosulfate consumed in the titration (mol/L); m is the weight of the sample (g); 162 is the molecular weight of glucose; 58 is the molecular weight of C_3H_6O; 29 and 9.67 are 1/2 and 1/6 of 58.

Through repeated experiments, the DS n which determined by above chemical method is significantly greater than the result determined by NMR method, and more closer to the actual value (Table 5.4) [30].

5.3.4.4.2. Spectrophotometry

This method is a kind of traditional method used in the DS determination of hydroxypropyl starch. It also can be used to determine the DS of HP-β-CDs. The principle is: first, hydroxypropyl of HP-β-CDs are hydrolyzed to propylene glycol. Then, propylene glycol is dehydrated to form propanol and propenyl. These two dehydration materials can react with triketohydrindene hydrate and produce purple complex. The quantity of hydroxypropyl can be calculated by

Table 5.4. DS results of HP-β-CDs [30].

Sample	Marked value	Value of chemical method	Value of NMR
1	14.0	13.82	12.95
2	7.0	6.91	6.64
3	1.0	0.94	0.89

spectrophotometry. This method is simple, and the experiment equipment is popular, but the operator must be skillful.

Triketohydrindene hydrate can be prepared by dissolving 0.15 g ninhydrin to 10 mL sodium hydrogensulfite (12.5%).

0.07 g sample was accurately weighted out and placed into a volumetric flask. 25 mL H_2SO_4 solution (0.5 mol/L) was added. Then, the volumetric flask was put in a boiling water bath until the sample dissolved. After cooling, it was then volumed with distilled water. The natural CD was treated with the same method. 1 mL sample, natural CD and 10, 30, 50, 70, 90, 110 mg/mL propylene glycol standard solutions were accurately measured into 25 mL colorimetric cylinders. The colorimetric cylinders were put in ice bath and 8 mL concentrated sulphuric acid was slowly added along the wall. Blend, and tighten the plug of colorimetric cylinders and accurately heat for 3 min in boiling water bath. Then, immediately put in ice bath until the temperature of the test liquid cooled to 5°C. 0.5 mL triketohydrindene hydrate was accurately measured and slowly added along the wall. Blend, and then put in 25°C water bath for 100 min for developing. Concentrated sulphuric acid was slowly added to 25 mL, slightly mix, and then put in 1 cm cuvette. Stand for 5 min, measure the absorbency at 59.5 nm. A standard curve was made between the propylene glycol concentration and absorbency. The value of propylene glycol conversion to hydroxypropyl was 0.7763.

5.3.4.4.3. NMR method

The principle is to use peak integral height of protons signal of 1 H NMR. Because the natural CDs do not contain methyl, hydroxypropyl content can be determined according to the integral height ratio of methyl in other CDs, and this is the DS.

5.3.4.4.4. MS method

HP-β-CD sample is permethylated, hydrolyzed and acetylated. The obtained acetylated D-glucitolethers can be analyzed by GC-MS. GC-MS were performed on a Finnigan Trace MS system with helium as the carrier gas [31, 32].

The steps of methylation are as follows. 10 mg of HP-β-CD is taken into a test tube with stopper, add 3 mL dimethylsulfoxide and mix the solution. Add 50 mg sodium hydroxide dry powder and 0.5 mL iodomethane, and then fill with nitrogen. Treat with ultrasonic at room temperature for 60 min. Add 5 mL water to suspend the reaction. Extract by 3 mL chloroform. Wash the organic phase twice with 5 mL water, then distill under reduced pressure to get faint yellow methylated product. Hydrolyze the product in 2 mol/L trifluoroacetic acid at 120°C for 1 h,

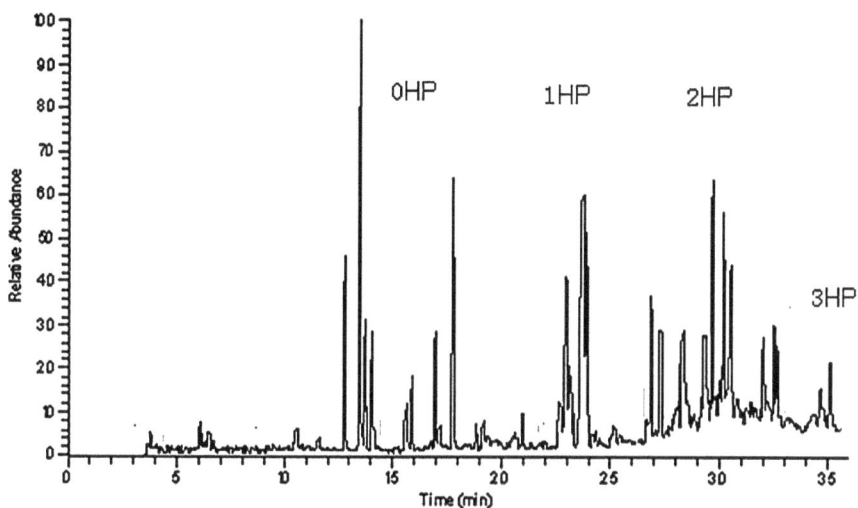

Fig. 5.7. GC-MS total ion current of acetylated D- glucose ester [32].

then dry by rotary evaporator under 40°C. Add 4 mL new sodium borohydride (0.5 mol/L, dissolved in 2 mol/L ammonia water), keep the reaction at 60°C for 60 min, and add 1 mL acetic acid to terminate the reaction. Dry and dissolve the residue in 1 mL methanol. Remove the residual boron by evaporation. Add 2 mL acetic anhydride, and keep at 100°C for 1 h. Cool in ice bath, add 1 mL ethanol, then evaporate to dry. Dissolve with 1 mL chloroform and refrigerate.

GC-MS conditions include: OV1701 capillary column (30 m×0.25 mm) with temperature programming. The experiment is initiated at 160°C and gradually the temperature is increased at the rate if 3°C/min until the temperature reaches 250°C. It is then allowed to stand for 5 min. The electron ionization (IE) is maintained at 70 eV.

Figure 5.7 shows the GC-MS total ion current of acetylated D-glucose ester. The order of peak was 0HP, 1HP, 2HP and 3HP.

According to the general rule of electron ionization mass spectrometry, acetylated D-glucose ester (Fig. 5.8-1) ionized by electron beam of ionization source mainly formed blow valuable pieces: a ($m/z = 117 + n2 \times 58$), b ($m/z = 233 + n3 + 6 \times 58$), c ($m/z = 161 + n2 + 3 \times 58$) and d ($m/z = 45 + n6 \times 58$).

Pieces a and b could determine the existing substituent and the place at C-2 or C-3/C-6. Piece c is weak, but there is an adjoint strong (c-60) peak, commonly used to judge the substitution site at C-3 or C-6. Piece d is very weak unless $n6 = 0$, and existing interference has little value. In addition, peaks that charge to mass ratio were 59 and 73 and very strong when hydroxypropyl existed (Fig. 5.8-3). There

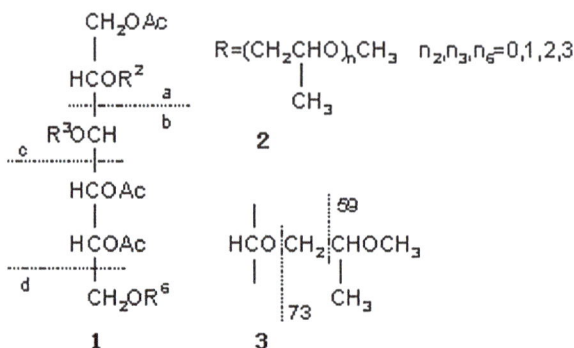

Fig. 5.8. Fragmentation of acetylated D-glucose ethers on MS [32].

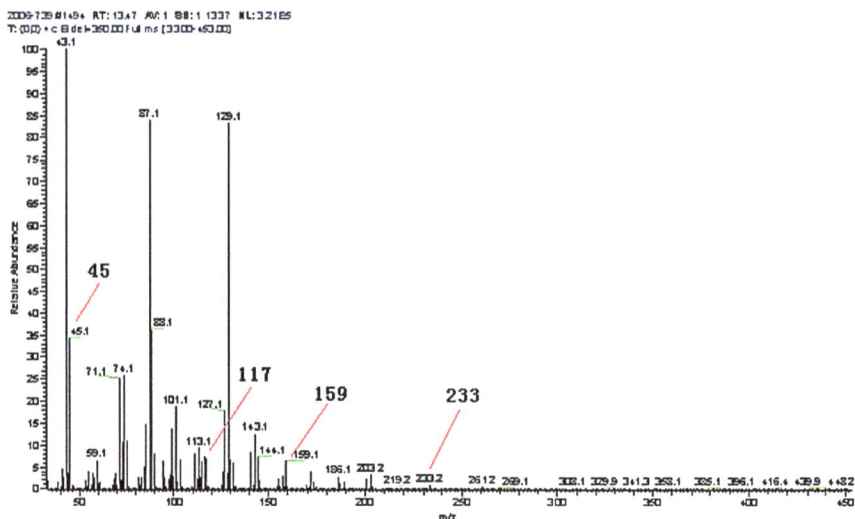

Fig. 5.9. Mass spectrum of unsubstituted acetylated D-glucose ethers.

should be a hydroxypropyl series connection when peaks that charge to mass ratio at 117 and 131 appear.

Based on the above analysis, total ion current of acetylated D-glucose ester and mass spectrogram of pieces are screened and calculated in detail. Mass spectrogram of pieces should have peaks a ($m/z = 117$), b ($m/z = 233$), c ($m/z = 161$) and d ($m/z = 45$) if the hydroxypropyl does not exist. Figure 5.9 is the typical mass spectrogram which does not have hydroxypropyl. From the total ion current, the peaks where the retention time was before 21 min were formed by unsubstituted

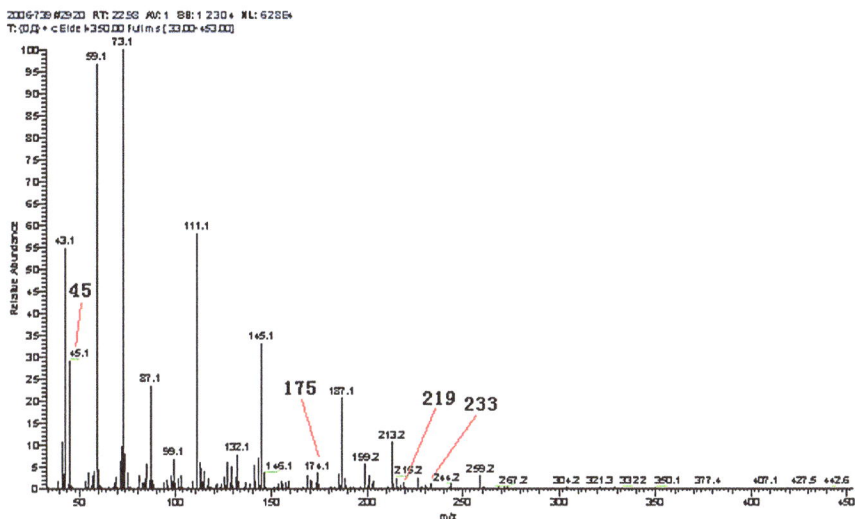

Fig. 5.10. Mass spectrum of monosubstituted acetylated D-glucose ethers.

glucose residues in HP-β-CDs. The peak where the retention time was between 21 min and 26 min were formed by monosubstituted glucose residues in HP-β-CDs. Now, it should have peak a ($m/z = 117$), b ($m/z = 291$), c ($m/z = 219$) and d ($m/z = 45$) if the hydroxypropyl substitution site is at C-3. It should have peak a ($m/z = 157$), b ($m/z = 233$), c ($m/z = 219$) and d ($m/z = 45$) if the hydroxypropyl substitution site is at C-2. It should have peak a ($m/z = 117$), b ($m/z = 291$), c ($m/z = 161$) and d ($m/z = 103$) if the hydroxypropyl substitution site is at C-6. Figure 5.10 shows the mass spectrum of monosubstituted acetylated D-glucose ethers. The hydroxypropyl substitution site is at C-2. Hydroxypropyl disubstitutions include three major types C-23, C-26 and C-36, and there are a few C-22'series connection disubstitution. It should have peak a ($m/z = 175$), b ($m/z = 291$), c ($m/z = 277$) and d ($m/z = 45$) if the hydroxypropyl substitution site is at C-23. It should have peak a ($m/z = 175$), b ($m/z = 291$), c ($m/z = 219$) and d ($m/z = 103$) if the hydroxypropyl substitution sites were at C-26. It should have peak a ($m/z = 117$), b ($m/z = 349$), c ($m/z = 219$) and d ($m/z = 103$) if the hydroxypropyl substitution site is at C-36. And it should have peak a ($m/z = 233$), b ($m/z = 233$), c ($m/z = 277$) and d ($m/z = 45$) if the hydroxypropyl substitution site is at C-22'. Figure 5.11 shows the mass spectrum of disubstituted acetylated D-glucose ethers. The hydroxypropyl substitution site is at C-23. From the total ion current, the peaks with retention time between 26 min and 34 min were formed by disubstituted

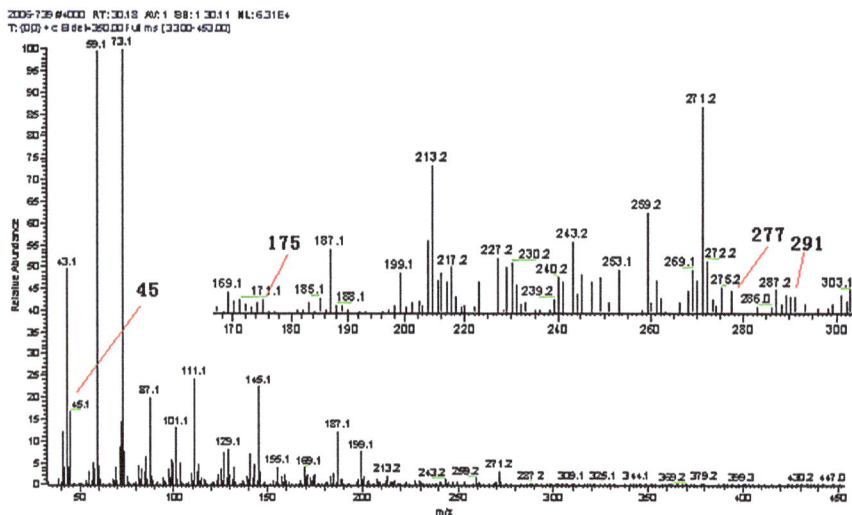

Fig. 5.11.   Mass spectrum of disubstituted acetylated D-glucose ethers.

glucose residues in HP-β-CDs. The peaks with retention time after 34 min were formed by trisubstituted glucose residues in HP-β-CDs. The common type of trisubstitution is C-236, and there are a few C-22'6 and C-266' trisubstitutions.

Table 5.5 shows the substituents distribution of acetylated D-glucose ester of HP-β-CD samples. Irrespective of high or low alkali concentrations, unsubstituted glucose units gradually reduce and substituted glucose units gradually increase with the increase of DS. Monosubstituted glucose units are major type in low DS samples, with a few disubstituted glucose units, but no trisubstitution. Along with the increase of DS samples, the amounts of monosubstituted glucose units and disubstituted glucose units both increased, but when disubstitution became the major type, a few trisubstitutions appeared. Weak alkaline conditions favor alkylation at more acidic C-2 hydroxyls (group A), while strong alkaline conditions favored alkylation at the more accessible C-6 hydroxyls (group B). The DS and ratio of alkylation at different points are calculated. In group A, the DS value of C-6 (DS (6)) is the highest, but C-2 and C-3 positions also had many substituents. The ratio of DS (2+3) to DS (6) is close to one. The distribution of substituents on primary (DS (6)) and secondary hydroxyl groups (DS (2+3)) was even, because the three free hydroxyl groups are all active under strong alkali conditions. The C-6 position has a little more opportunity to be substituted due to steric hindrance. Different from this group, the substituents concentrate at C-2 in group B. The DS value of secondary hydroxyl groups is about three times of primary face.

Table 5.5. Distribution of substituents and the respective DS.

Substitution site	Group A				Group B				Standard
	1	2	3	4	5	6	7	8	9
–	67.38	58.81	41.38	20.12	67.32	49.51	37.94	18.29	48.20
3	0.83	2.81	6.31	4.45	2.57	1.67	1.50	1.83	2.09
2	5.77	9.75	13.11	12.21	16.95	24.53	21.29	29.37	20.54
6	12.79	13.92	14.09	20.52	1.09	3.58	2.35	4.15	3.67
23	0.59	1.17	0	1.25	3.85	10.76	15.78	18.64	11.00
36	1.22	2.69	1.81	6.78	6.16	3.89	8.31	6.01	6.63
22′	0	0	0	0	0.61	1.74	1.60	2.19	2.02
26	11.39	10.36	22.84	31.24	0.73	2.03	5.45	13.23	2.72
66′	0	0	0.49	1.83	0.74	1.22	2.02	1.77	1.62
236	0	0.49	0	1.51	0	0.49	1.57	1.86	0.71
22′6	0	0	0	0	0	0.59	2.16	2.65	0.86
266′	0	0	0	0.12	0	0	0	0	0
Unsubstituted	67.38	58.81	41.38	20.12	67.32	49.51	37.94	18.29	48.20
$\sum$Mono-substituted	19.39	26.48	33.51	37.18	20.61	29.78	25.14	35.35	26.3
$\sum$Disubstituted	13.2	14.22	25.14	41.1	12.09	19.64	33.16	41.84	23.99
$\sum$Trisubstituted	0	0.49	0	1.63	0	1.08	3.73	4.51	1.57
DS(3)	0.18	0.50	0.57	0.98	0.88	1.18	1.90	1.98	1.43
DS(2)	1.24	1.53	2.51	3.24	1.59	2.97	3.61	5.10	2.85
DS(2+3)	1.43	2.03	3.08	4.22	2.47	4.15	5.51	7.08	4.28
DS(6)	1.78	1.92	2.78	4.48	0.66	0.91	1.67	2.20	1.25
DS(2+3)/ DS(6)	0.80	1.05	1.11	0.94	3.73	4.55	3.30	3.22	3.43
Total DS	3.21	3.95	5.77	8.68	3.14	5.06	7.19	9.27	5.53
Molecular weight	1321	1364	1470	1638	1317	1428	1551	1673	1456

5.4. Preparation and Analysis of CDP

5.4.1. *CDP*

CDPs are the high molecular weight CD derivatives which contain several CD units. CDPs keep the ability of CD, for example, encapsulation, control-released, catalysis and recognition. Furthermore, they have the characteristics of superpolymer such as high mechanical strength, good stability and chemical adjustability. The solubility is also improved. Therefore, CDP have special functions and wide applications in molecular recognition, chromatographic separation, environmental pollution control, medicine and food industry. More and more studies are focused on the CDP. In 1965, Solms first reported the synthesis of the CDP. In the following ten years, many investigations on the CDPs were carried out and some new synthesis methods were developed. Since the 1980s, along with the development of CDP research, the applications of CDP become broader [33].

According to differences in water-solubility, CDPs can be divided into water-soluble, water insoluble and water swelling CDPs. Water insoluble CDPs have applications in separation and analysis technology, trace organic capture and special material because CDPs can be easily separated and are recyclable. Most CDPs belong to water insoluble polymer. Some other water-soluble CDPs, especially bonding to native polymer chain, may become the new materials and carriers for drug development formulations.

5.4.2. *Preparation of CDP*

The preparation methods of CDP include: crosslinking, solid load, blending and a combination of crosslinking and solid load, etc. Several typical examples of CDP preparations and applications are introduced below in detail.

5.4.2.1. *Crosslinked CDP*

5.4.2.1.1. Water insoluble CD/EPI polymers

Weigh defined amount of β-CD and 20% sodium hydroxide solution into a triangle flask and agitate until dissolved completely at 60°C. Drop 30 mL EPI into the solution while stirring and keep the reaction to gain a hard gel. Take out the gel and wash it with water and acetone until no chloride ions remain. Filter the lotion, then dry at 60°C for 48 h under vacuum and grind afterwards. The white powder is the crosslinked polymer product.

Mix defined amount of β-CD, starch and 20% sodium hydroxide solution into a triangle flask, stir in the hot water bath for 24 h. Rest of the steps are the same as mentioned above. A gray crosslinked polymer will be gained.

The melting point of the crosslinked polymer is 271°C and 272°C. Put 1.0 g polymer into 50 mL beaker, then add 50 mL water, 0.1 mol/L HCl, 0.5 mol/L H_2SO_4, 0.1 mol/L NaOH, 4 mol/L NaOH, and organic solvent chloroform, ethanol and acetone, respectively, stir for 30 min. It could be found that the crosslinked polymer is insoluble in water when the common acid, alkali and organic solvents are at room temperature.

This insoluble CDP can be used as a stationary phase in the capillary column for the separation of isomers. The preparation method is as follows: the elastic quartz capillary column is washed with 10 mL methanol, blow dried, then purged with N$_2$ at 250°C for 4 h. OV-17 is coated using super dynamic method at room temperature under 810 kPa, blow dried with N$_2$, aged for 2 h at 160°C. At certain temperature and pressure, the CDP stationary liquid (benzene : isopropyl ether : methylene chloride = 2:2:1 (v:v:v)) is coated rapidly using the back super dynamic method, blow dry with N$_2$, age 1 h at 60°C. Then the temperature rises upto 180°C at the rate of 2°C/min, and aged for 6 h again [34].

CDP cannot only separate the compounds with different boiling point and polarity, but can also easily separate some similar aromatic hydrocarbons, especially chiral compounds, because of its unique cavity structure. This modified quartz capillary column has very strong chromatographic separation ability and is a useful high selectivity capillary column.

5.4.2.1.2. Water soluble CD/EPI polymers

Dissolve 9.0 g β-CD in 30 mL NaOH aqueous solution (20%), add 120 mL distilled water and fully mix the solution. Drop 7.0 g EPI slowly at 60°C within 40 min, and then adjust the temperature to 65°C. Keep the reaction for 24 h. Then neutralize the reaction mixture with 2 mol/L HCl and solidify in five days under low temperature to make the residual EPI and insoluble polymer form solid gel. Filter, remove the salt and residual β-CD. Soluble β-CDP can be gained; production rate is about 25%.

Water soluble β-CD crosslinked polymer is one kind of macromolecular substance which has the chain structure. It has high solubility in water. The polymer has special stereoscopic selection to line molecules which contain double benzene ring. It can make the guest more soluble, more stable and more susceptible in fluorescence spectrum. The cavity of β-CD can increase the density of electron cloud of guest molecules. Fluorescence emission of no fluorescence and weak fluorescent molecules enhance markedly under the synergy of multi cavity of β-CD in the polymer. Especially, the double benzene ring interval is 3–4 molecules and the fluorescent launching ability increase apparently. The result shows that water soluble β-CD crosslinked polymer might become a new material used in fluorescence identification of different double guest compounds [35].

5.4.2.1.3. Water insoluble CD/toluene diisocyanate polymers

β-CD is scattered in DMSO, dissolved at 70°C with stirring. Added toluene diisocyanate (TDI) (β-CD : TDI = 7:9) and keep the reaction for 2–6 h. Then add acetone to terminate the reaction. The obtained solid product is porphyrized and washed with boiling water and hot ethanol respectively. The powder is dried under vacuum at 60°C for 48 h. A gray granular CDP is gained. The yield is about 98% [36].

5.4.2.1.4. β-CDP microspheres

First, β-CD is dissolved in 20% NaOH aqueous solution, then EPI is dropped in with stirring. The mixture reacts for 1.5 h at room temperature, and then 200# kerosene which contains emulsifier (n (tween 20) : n (span 80) = 1:3) is added and vigorously mixed for 5 min to make the water phase disperse. Finally, the reaction system is heated up to 60°C, stirring speed is reduced while continuing to react for 8 h. The product is filtered and washed with diluted hydrochloric acid, methanol, distilled water and acetone, respectively. After wash, the product is dried under vacuum.

At present, CDP which is used in medication control release is a hot research direction in CD science. Relevant reports are limited, especially the study on control release of traditional Chinese medicine. Researches in which β-CDP is used as microspheres to control the release of medicine are important for development of new form and modernization of traditional Chinese medicine.

Some researchers prepared magnetic polymer microsphere using β-CD which included magnetic material by crosslinking method [37].

6 mL 0.5 mol/L of Fe2 + solution and 10 mL 0.5 mol/L of Fe3 + solution is mixed, 60 mL 3 mol/L NaOH aqueous solution is added and allowed to react for 30 min. Then, 50 mL 4% OP emulsifier solution is added. The solution continued to react for 30 min and gained the scattered magnetic liquid. The magnetic liquid is allowed to stand and supernatant liquor is removed. 20 mL 4% OP solution is added then treated with ultrasonic for 30 min. 4% of OP solution is added diluting the solution to 75 mL. The solution is heated to 50°C. 8 g β-CD is dissolved in 20 mL 20% NaOH aqueous solution, and then mixed with the above 75 mL magnetic liquid. The mixture is treated with ultrasonic for 10 min. EPI is dropped in with stirring at 50°C water bath, and allowed react for 1 h. After reation, the mixture is kept for 3–4 h, remove supernatant liquor, the magnetic microspheres are gained, filtered, washed, dried and then the solid product is gained.

Magnetic polymer microsphere is a kind of microspheres which combines inorganic polymer and magnetic materials. It is magnetic and surface modifiable, therefore, has a wide application in the targeted drugs, immobilized enzyme and cell separation, etc.

5.4.2.1.5. Nanotube CDPs

Rotaxane and polyrotaxane are a kind of novel supramolecules which are formed by ring molecules and linear polymers through complexation. Some researchers focus on them because generation of this kind of supramolecules is not dependent on the covalent bond and so, they develop very fast. Nanotube CDP can be prepared by CD polyrotaxane intermediate. This polymer shows special selectivity to different macromolecules.

To prepare nanotube CDPs, CD is first reacte with polyethylene glycol (PEG) where, both ends of amino are replaced to form analog polyrotaxane. PEG chain is threaded through the CDs. Then 2, 4-dinitrofluorobenzene is reacted with amino on the ends of PEG chain to prevent CD falling off from the PEG chain. CD polyrotaxane is thus formed. Then, the adjacent CDs are crosslinked, and the barrier 2, 4-nitrobenzene is taken off, the PEG chain is removed, and the nanotube CDP is obtained.

5.4.2.2. *Immobilized CDP*

The carriers of chemically immobilized CDP basically are of three kinds: inorganic polymer (such as silica gel), natural polymer (such as chitosan, cellulose, etc.) and synthetic polymer (such as polystyrene microspheres, etc.) [38, 39].

5.4.2.2.1. Immobilized to silicon

20 g activated silica gel is scattered in 300 mL anhydrous toluene, 8 mL γ-chloropropyltriethoxysilane is added and allowed to react for 8 h under nitrogen protection at 110°C. Then it is cooled down to room temperature. Solid product (CD-SiO$_2$) is filtered and washed with DMF, toluene and water respectively, dried under vacuum for 48 h. 5 g dried β-CD is dissolved in 50 mL anhydrous DMF. 0.5 g NaH is added, and allowed to react for 0.5 h. The insoluble materials are filtered, SiO2–Cl is added in filtrate, slowly warmed to 90°C and reacted for 24 h. The product (CD-SiO$_2$) is filtered and washed with DMF, and water in turn, dried under vacuum for 48 h, and the solid product is thus gained.

Some other methods are immobilized by amino bonds. For example, 6-chlorpromazine-CD and single-6-sulfonic acid ester-CD are used as intermediates. Silica gel carrier with the amino groups is used to immobilize CDs to obtain CD silica gel bonded phase.

Inorganic silicon carriers have high mechanical strength, but alkali resistance is poor, the adsorption ability is low. Nonspecific adsorption of silicon hydroxyl is strong. An organic layer is bonded on the silicon which can passivate silicon hydroxyl and create a gentle contact surface. It is expected to prepare an adsorbent with both high adsorption and high mechanical properties. Especially, the cavity structure of CDs is very effective for the separation of isomers.

5.4.2.2.2. Immobilized to filter paper

Filter paper is cut in a circle, about 4 cm in diameter and soaked in pure water for 0.5 h, then soaked with 5 mol/L NaOH solution for 0.5 h. The filter paper is washed with pure water, then put in a conical flask which contains mixture of NaOH : dimethyl sulfone : EPI 2 : 4: 5, volume ratio. The conical flask is put in 50°C water bath for 1 h for activation. Then the mixture is poured out, the activated filter paper is washed with pure water and put into 0.3 mol/L β-CD NaOH solution. This is allowed to react for 5 h at 50°C, and washed with pure water until the pH of cleaning solution is about 7. Finally, this is soaked with acetone and dried in the air.

5.4.2.2.3. Immobilized to cellulose

At present, the main chemical crosslinking methods used in immobilization of CD to cellulose are EPI method, hydroxymethyl acrylamide method, monochlortriazingl method poly carboxylic acid method, etc.

1 g viscose acetal fiber is put in 250 mL conical flask. To this, 25 mL distilled water and 4 mL 0.1 mol/L NaOH solution are added and allowed to swell for 1 h. 55 mL distilled water and 7 mL EPI and 6 mL 40% NaOH solution are added in the conical flask. The conical flask is put in a shaking water bath for reacting. After the reaction, distilled water is added to wash the product to neutraize. Residual EPI is detected by sodium thiosulfate and phenolphthalein indicator. Until there is no EPI in the product, this filtered and washed with acetone two times, then dried under 50°C. The epoxidation cellulose fiber is put into 40% sodium hydroxide β-CD reaction solution (the quality ratio of epoxidation cellulose fiber and alkali is 1:50, epoxidation cellulose fiber and β-CD is 1:1.1), reaction temperature is kept at 45°C, and reacted for 2.5 h. Distilled water is added to stop the reaction. The product is washed and filtered until the wash water becomes neutral. The product is collected and dried at 50°C.

5.4.2.2.4. Immobilized to chitosan

Figure 5.12 shows the synthesis route of CD immobilized to chitosan. 4.0 mL EPI, 0.08 mL perchloric acid, 0.5 mL ethanol and 0.5 mL water are mixed. 1.0 g chitosan and 60 mL toluene are added and reacted for 4 h at 90°C, filter. The filter cake is washed with acetone (2 mL×15 mL), dried at 60°C, and the white pulverous N-(3-chloro-2-hydroxyl)-propylchitosan is gained. 0.6 g N-(3-chloro-2-hydroxyl)-propylchitosan, 0.8 g β-CD, 1.0 g anhydrous sodium carbonate and 30 mL distilled water are mixed well and reacted for 2 h under reflux at about

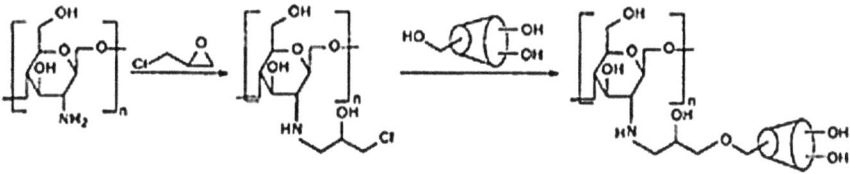

Fig. 5.12. Synthesis route of CD immobilized chitosan [40].

90°C. Filter and filter cake are washed to neutralize with 10 mL 0.01 mol/L in NaOH, 5 mL 0.01 mol/L HCl, 50 mL distilled water respectively, then washed with 15 mL ethanol and 10 ml acetone. The residual product is dried at 60°C to gain the buff pulverous immobilized chitosan [40].

CD/chitosan polymer can be prepared using EPI as the crosslinking agent under the protection of benzaldehyde on amino. This polymer retained the free amino of chitosan and added the hydrophobic cavity of CD. Adsorbability is superior to chitosan.

The grafted agent also can use glutaraldehyde, etc. Furthermore, CD intermediates such as CD sulfonic acid ester can be used in immobilization.

5.4.2.2.5. Immobilized to starch

β-CD is dissolved in NaOH solution with certain concentration at room temperature. EPI is slowly dropped in, then they are mixed for 0.5 h. The gelatinized starch is added in the reaction system; continue to stir for 2 h, then cooled. Absolute ethyl alcohol is added in for precipitation. Separate the sediment by centrifugal force. The sediment is dispersed with a small amount of distilled water, neutralized with diluted hydrochloric acid and precipitated with absolute ethyl alcohol. The sediment is separated again by centrifugal force. The sediment is washed with 50% ethanol to remove the remnants of CD and sodium chloride and dried under vacuum at 75°C.

5.4.2.2.6. Immobilized to polyacrylamide

4.30 g polyacrylamide is added in 160 mL distilled water, mixed in a 50°C water bath for dissolving. 2.0 g β-CDs-sulfonic ester is added by bath to react in 50°C water bath for 24 h. The solvent is removed through evaporation and the solid product is washed by methanol and ether respectively. Then, it is vacuum dried and white solid polymer products are obtained.

CD immobilization which uses organic synthetic polymer as carrier is a novel technology and relevant reports appeared until the 1970s. This kind of carrier is structure adjustable with good mechanical performance. Thus, they are more practical.

5.4.2.2.7. Immobilized to crosslinked chloromethylate polystyrene

5.0 g crosslinked chloromethylate polystyrene (CMPS, degree of crosslinking 7) is added in a 250 mL three-necked flask. 30 mL cosolvent DMF which is dried by anhydrous sodium sulfate is added and allowed to swell for 10 h. Then, β-CD which dissolved in DMF and sodium hydroxide are added in and allowed to react for 24 h under backflow at 45°C. The reactants are leached and washed with distilled water. By vacuum drying under room temperature, we gain the immobilized CD/CMPS polymer. Through the test, it can be found that β-CD is immobilized to CMPS by ether bond in alkaline medium. The immobilization reaction occurs between chloromethyl of the carrier and the hydroxyl of β-CD.

Compared with the carrier, the immobilization is better at hydrophilicity, chemical stability and thermal stability aspects. This polymer can be used in wastewater treatment which can remove metal ions and organic pollutants at the same time and can be reused through deabsorption.

5.4.2.3. *Polymer blend*

CDs are equably dispersed in polymer. A kind of soluble polymers (polyvinyl acetate, PAM) is chosen and dissolved in solvents (acetone, ethyl acetate). Then CD is added and scattered in the solvents. Finally, the solvent is evaporated and the CDP blend is prepared. It is the gentlest CD immobilization method, as it is not a chemical reaction, and is widely used as the operation is simple. For example, the polymer blend of CD/polyacrylamide can be used as filling materials of capillary chromatographic column. Its separation efficiency is as high as 10,000 layer/15 cm. If spices or fungicide are added to CD before blending with polyethylene, the blend products can be used as flavor enhancer or antibacterial packaging material.

5.4.2.4. *CD/macromolecule inclusion complex*

CDs can also form inclusion complex with the long chain high polymer. The molecular structure of the complex is that a linear polymer long chain "axis" threads a lot of (20 to 40) ring CD molecules, namely CDs like "rotor". Chemists called it polyrotaxanes. Although it is not CDP that connected with chemical bonds, its special structure differs from CD and the carrier shows the selectivity recognition to different polymers and has special performance. It has already caused great interest to researchers. Harada's group (Japan) is one of the earliest groups who began to study CD polyrotaxanes [41]. They put PEG with a molecular weight between 400–1,000 in α-CD saturated solution and found that the solution became cloudy. The sediment which is separated by centrifugal or filtration was confirmed as inclusion complex formed between CD and PEG. Harada found that the α-CD/PEG complex was "tunnel" type crystal according to the X-ray figure of the complex. The molecular dynamics simulation of the polyrotaxanes which formed between CD and PEG and PPG further proves that the structure of the complex was "tunnel" type.

5.4.3. *Analysis of CDP*

5.4.3.1. *Infrared spectrum*

The common method used in CDP characterization is infrared spectrum. The formation and link type of CDP are determined through contrasting the infrared spectrum change between CD and its polymerization product. For example, infrared spectrum of crosslinked CDP should appear as absorption peaks of crosslinking agent groups except absorption peaks of CD. Figure 5.13 is the infrared spectrum of crosslinked CDP which as formed by β-CD and EPI. Compared with β-CD, stretching vibration peak at $3,440\,\text{cm}^{-1}$ became narrow, indicating that the crosslinking agent had reacted with hydroxy on CD, and reduced the number of freehydroxy. It thus opened the intramolecular hydrogen bond. C–O absorption peak at $1,050\,\text{cm}^{-1}$ was enhanced proving that etherification had happened.

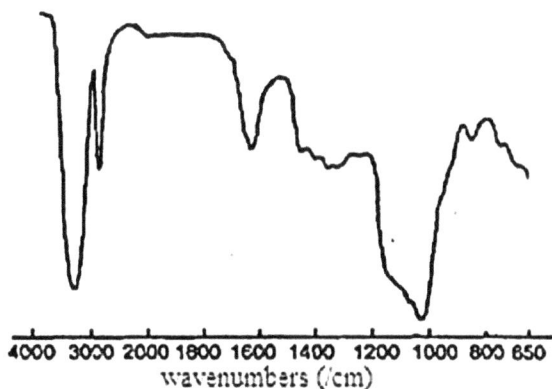

Fig. 5.13. Infrared spectrum of crosslinked CDP [40].

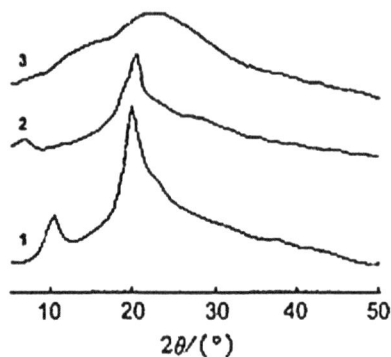

Fig. 5.14.   X-ray spectrum of chitosan (1) middle product (2) and final product (3) [40].

5.4.3.2. *X-ray*

Some polymers can crystallize, and the change in the structure can be found by comparing the difference of X-ray before and after polymerization. The X-ray spectrum of chitosan (1) middle product (2) and final product (3) is shown in Fig. 5.14.

Chitosan had strong diffraction absorption peaks at 10° and 20°. In the spectrum of middle product N-(3-chloro-2-hydroxyl)-propylchitosan, the 10° characteristics absorption peaks had disappeared, and the 20° peak had weakened. This was because the amino in the chitosan molecules reacted with EPI and formed an intermediate which contained chlorinated hydroxypropyl, resulting in weakened crystallinity. The 20° peak of immobilized chitosan (3) was further weakened compared with (2) because it was the grafted product of N-(3-chloro-2-hydroxyl)-propylchitosan with CD and had a further lower crystallinity [40].

5.4.3.3. *Scanning electron microscope*

Scanning electron microscopy (SEM) can visually observe the appearance and structure of tested material, determining the change caused in polymerization. For example, Fig. 5.15 is an SEM photo of CD grafted cellulose. It can be found that nature cellulose (1) was smooth and the texture was clear. The cellulose only treated with grafting agent polyacrylic acid (2) was rough and the texture was not clear; CD grafted cellulose (3, 4) had nonuniform grafted CD, but the surface texture changed a little. It indicated that cellulose surface was damaged when treated and with acrylic acid. However, the added CD could join and participate in the esterification reaction; reduce contact and reaction between the polyacrylic acid and cellulose. So that part of the cellulose surface remained unchanged. Therefore,

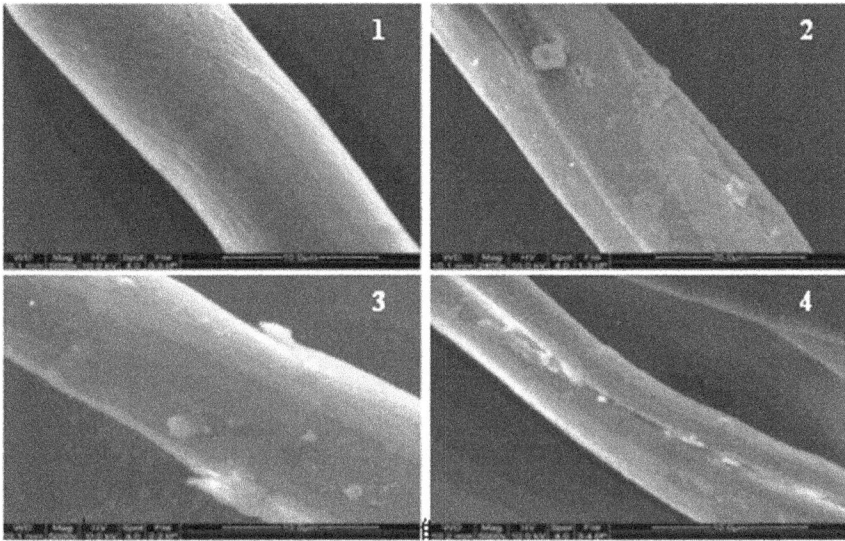

Fig. 5.15. SEM photos of cellulose and CD grafted cellulose.1: nature cellulose, 2: cellulose treated with grafting agent polyacrylic acid, 3, 4: CD grafted cellulose [42].

CD grafted not only connected the functional groups for cellulose, but also did not excessively change the cellulose surface structure [37].

5.4.3.4. *Determination of CD content in CDP*

5.4.3.4.1. Phenol colorimetric method

The CDP is hydrolyzed by sulfuric acid and the gained CD is colored with phenol and the CD content is determined by spectrophotometer. This method is suitable for most of the CDPs except that the substrate is starch or the similar polysaccharides. The detailed operation is as follows:

Different concentration glucose standard solutions (1 mL) is pipetted into the test tube with plug. 1 mL distilled water, 1 mL phenol solution (8%) and 5 mL of sulfuric acid were added respectively, fully mixed, allowed to stand for 10 min and vibrated. Then it was put in water (25°C to 30°C) for 20 min. The solutions shows gradually deepened orange color according to the concentration from low to high. Absorbency value was determined and standard curves were drawn.

By accurately weighing a certain amount of CDP with the same operation method, we can determine the absorbency value and calculate the CD content according to the standard curve.

The polymer degree of crosslinking or apparent immobilization quantity can be calculated according to the calculated CD content.

5.4.3.4.2. Phenolphthalein colorimetric method

The principle is that the absorbency of phenolphthalein can be reduced after phenolphthalein is embedded by CD and absorbency reduction has linear relation with CD content in certain concentration range. Apparent CD can be calculated through the content of phenolphthalein which is embedded by CD.

A series of CD standard solutions (concentration range from 0 to 30 μg/mL) are prepared and added to phenolphthalein standard solution (pH $= 10.8$), and oscillated for 30 min. Absorbency value is determined at 552 nm and then the working curve is drawn. An accurate amount of CDP sample is taken, added to the above standard solution and oscillated for 30 min. It is filtered to remove insoluble polymer. The absorbency value is determined at 552 nm. The CD content is calculated on the basis of the working curve.

The determination result is on the high side because the CDP net structure does not powers adsorption ability. For some polymer, the immobilized CD can be taken off through alkali elution, and then colorimetric assay.

5.4.3.4.3. Gravimetric method

Gravimetric method determines the content of CD by precisely measureing the weight change of the sample before and after immobilization. This method is based on different content of CD which when immobilized on the carrier has no effect on quantity of crosslinking agent which was grafted on the carrier. The determination result is not very accurate and is suitable for applications which have low expectations on quantitative analysis.

References

1. Del Valle, EMM (2004). Cyclodextrins and their uses: A review. *Process Biochemistry*, 39(9), 1033–1046.
2. Tong, LH (2001). *Cyclodextrin Chemistry*. Beijing; Science Press, (in Chinese).
3. Cucinotta, V, F D'Alessandro, G Impellizzeri *et al.* (1992). Synthesis and conformation of dihistamine derivatives of cyclomaltoheptaose (β-cyclodextrin). *Carbohydrate Research*, 224, 95–102.
4. Takeo, K (1990). *Carbohydrate Research*, 200, 481–487.
5. Fenichel, L, P Bako, L Toke *et al.* (1988). In *Proceedings of 4th International Symposium on Cyclodextrins*, p. 113 Munish.

6. Ashton, PR, SE Boyd, G Gattuso *et al.* (1995). A novel approach to the synthesis of some chemically-modified cyclodextrins. *The Journal of Organic Chemistry*, 60(12), 3898—3903.

7. Hao, AY and LH Tong (1995). Preparation and NMR characterization of mono-2-O-(2-hydroxypropyl)-β-cyclodextrin. *Chinese Journal of Synthetic Chemistry*, 3(4), 369–371 (in Chinese).

8. Fügedi, P (1989). Synthesis of heptakis (6-O-tert-butyldimethylsilyl) cyclomalto-heptaose and octakis (6-O-tert-butyldimethylsilyl) cyclomaltooctaose. *Carbohydrate Research*, 193, 366.

9. Luna, EA, DG Vander Velde, RJ Tait *et al.* (1997). Isolation and characterization by NMR spectroscopy of three monosubstituted 4-sulfobutyl ether derivatives of cyclomaltoheptaose (β-cyclodextrin). *Carbohydrate Research*, 299(3), 111–118.

10. Jullien, L, J Canceill, B Valeur *et al.* (1996). Multichromophoric cyclodextrins. 4. Light conversion by antenna effect. *Journal of the American Chemical Society*, 118 (23), 5432–5442.

11. Bonomo, RP and V Cucinotta (1991). Conformational features and coordination properties of functionalized cyclodextrins. Formation, stability, and structure of proton and copper(II) complexes of histamine-bearing beta-cyclodextrin in aqueous solution. *Inorganic Chemistry*, 30(13), 2708–2713.

12. Murakami, T, K Harata and S Morimoto (1987). Regioselective sulfonation of a secondary hydroxyl group of cyclodextrins. *Tetrahedron Letters*, 28(3), 321–324.

13. Zhanga P and AW Colemana (1993). Synthetic route for selective modification of the secondary hydroxyl face of cyclodextrins. *Supramolecular Chemistry*, 2(4), 255–263.

14. Fujita, K, S Nagamura and T Imoto (1984). Convenient preparation and effective separation of the C-2 and C-3 tosylates of α-cyclodextrin. *Tetrahedron Letters*, 25(49), 5673–5676.

15. Fujita, K, T Tahara, T Imoto *et al.* (1986). Regiospecific sulfonation onto C-3 hydroxyls of .beta.-cyclodextrin. Preparation and enzyme-based structural assignment of 3A, 3C and 3A3D disulfonates. *Journal of the American Chemical Society*, 108(8), 2030–2034.

16. Hao, AY, LH Tong and FS Zhang (1995). *Carbohydrate Research*, 277, 333.

17. Ni, W and J Bai (2005). Synthesis and application of cyclodextrin polymer. *New Chemical Materials*, 9, 60–62 (in Chinese).

18. Schomburg, G, A Deege and H Hinrichs (1992). Preparation, purification, and analysis of alkylated cyclodextrins. *Journal of High Resolution Chromatography*, 15(9), 579–584.

19. Szejtli, J, A Lipták and I Jodál (1980). Synthesis and 13C-NMR spectroscopy of methylated beta-cyclodextrins. *Starch*, 32(5), 165–169.

20. Szente, L and I Szejtli (1999). Highly soluble cyclodextrin derivatives: Chemistry, properties, and trends in development. *Advanced Drug Delivery Reviews*, 36(1), 17–28.

21. Rao, CT, HM Fales and J Pitha (1990). Pharmaceutical usefulness of hydroxypropy-lcyclodextrins: "e pluribus unum" is an essential feature. *Pharmaceutial Research*, 7, 612–615.

22. Dodziuk, H (2002). Rigidity versus flexibility. A review of experimental and theoretical study pertaining to the cyclodextrin nonrigidity. *Journal of Molecular Structure*, 614, 33–45.

23. Gould, S and RC Scott (2005). 2-Hydroxypropyl-β-cyclodextrin (HP-β-CD): A toxicology review. *Food and Chemical Toxicology*, 43(10), 1451–1459.

24. Pitha, J, J Milecki, H Fales, L Pannell *et al.* (1986). Hydroxypropyl-β-cyclodextrin: Preparation and characterization; effects on solubility of drugs. *International Journal of Pharmaceutics*, 29(1), 73–82.

25. Yuan, C, BG Liu and CG Chen (2010). Optimization of preparation process of hydroxypropyl-beta-cyclodextrin by response surface methodology. *Proc. 2010 International Conference on Challenges in Environmental Science and Computer Engineering*, pp. 26–28. CESCE.

26. Irie, T, K Fukunaga, A Yoshida *et al.* (1988). Amorphous water-soluble cyclodextrin derivatives: 2-hydroxyethyl, 3-hydroxypropyl, 2-hydroxyisobutyl, and carboxamidomethyl derivatives of β-cyclodextrin. *Pharmaceutical Research*, 5(11), 713–716.

27. Yuan, C and ZY Jin (2008). Preparation and stability of the inclusion complex of astaxanthin with hydroxypropyl-beta-cyclodextrin. *Food Chemistry*, 109(2), 264–268.

28. Kohata, S, K Jyodoi and A Ohyoshi (1993). Thermal decomposition of cyclodextrins (α-, β-, γ-, and modified β-CyD) and of metal-(β-CyD) complexes in the solid phase. *Thermochimica Acta*, 217, 187–198.

29. Cheng, SZD, CY Li, BH Calhoun *et al.* (2000). Thermal analysis: The next two decades. *Thermochimica Acta*, 35(1), 59–68.

30. Hao, AY and HG Zhang (2002). Substitute degree determination of hydroxypropyl-β-cyclodextrin by chemical method. *Journal of Shandong University (Natural Science)*, 37(2), 52–55 (in Chinese).

31. Pitha, J and TC Rao (1990). Distribution of substituents in 2-hydroxypropyl ethers of cyclomaltoheptaose. *Carbohydrate Research*, 200, 429–435.

32. Yuan, C, ZY Jin and XH Li (2008). Evaluation of complex forming ability of hydroxypropyl-β-cyclodextrins. *Food Chemistry*, 106, 50–55.

33. Fenyvesi, É (1988). Cyclodextrin polymers in the pharmaceutical industry. *Journal of Inclusion Phenomena and Macrocyclic Chemistry*, 6(5), 537–545.

34. Yi, J and K Tang (2000). β-Cyclodextrin cross-linked oligopolymer used as a capillary gas chromatographic stationary phase and separation of enantiomers and isomers. *Chinese Journal of Analytical Chemistry*, 28(10), 1291–1294 (in Chinese).

35. Su, X, L Liu and H Shen (1997). Studies on the fuluorimetric molecular recognition of Schiff bases containing double aromatic guests with water soluble β-cyclodextrin cross-linking polymer. *Chemical Journal of Chinese Universities*, 18(8), 1275–1280 (in Chinese).

36. Rohrbach, RP (1990). Method of preparing cyclodextrin-coated surfaces. US Patent: 4917956.

37. Lv, Ha and Z Liu (2005). Preparation and characterization of magnetic β-cyclodextrin polymer microspheres. *Journal of Guilin University of Technology*, 25(4), 543–547 (in Chinese).

38. Ruderisch, A, J Pfeiffer and V Schurig (2003). Mixed chiral stationary phase containing modified resorcinarene and β-cyclodextrin selectors bonded to a polysiloxane for enantioselective gas chromatography. *Journal of Chromatography A*, 994(2), 127–135.

39. Ni, W and J Bai (2005). Synthesis and application of cyclodextrin polymer. *New Chemical Materials*, 33(9), 60–63 (in Chinese).

40. Yi, Y and Y Wang (2005). Preparation and characterization of chitosan-g-β-cyclodextrins. *Chinese Journal of Synthetic Chemistry*, 13(2), 180–182 (in Chinese).

41. Harada, A, M Furue and S Nozakura (1976). Cyclodextrin-containing polymers. 1. Preparation of polymers. *Macromolecules*, 9(5), 701–704.

42. Jin, Z, X Xu and H Chen (2009). *Cyclodextrin Chemistry*, Beijing: Chemical Industry Press. pp. 275–276. (in Chinese).

6

BASIC APPLICATION OF CYCLODEXTRINS IN SUPERMOLECULE CHEMISTRY

Tao Feng*, Ai-Quan Jiao† and Zheng-Yu Jin‡

*School of Perfume and Aroma Technology,
Shanghai Institute of Technology,
Shanghai 200235, China
†School of Food Science and Technology,
Jiangnan University Wuxi 214122, China
‡The State Key Laboratory of Food Science and Technology,
School of Food Science and Technology, Jiangnan University,
Wuxi 214122, China

6.1. Overview of Supermolecule Chemistry

6.1.1. *Definition and research area of supermolecule chemistry*

Supermolecule chemistry, including molecular recognition, molecular catalysis, transmission procedure and carrier design, molecule and supermolecule device and molecule self-assembly, is a kind of newly developed interdisciplinary subject in the recent 30 years [1]. Due to its wide applications in industry, information, life and material science etc, supermolecule chemistry has been receiving increased attention [2].

6.1.2. *The status of cyclodextrin in supermolecule chemistry*

Supermolecule interaction is a kind of intermolecular interaction with molecule recognition ability. The interaction includes van der Waals, electrostatic attraction, hydrogen bonding, π interaction and hydrophobic interaction, etc under the steric effect. Such an interaction is similar to the intermolecular specific combination as the "lock and key" theory, which is the basic of supermolecule structure formation. Because cyclodextrin (CD) has a special cone structure, i.e. the inside of its cavity is hydrophobic whereas the outside is hydrophilic, it can form inclusion complex with many organic or inorganic molecules through van der Waals, hydrophobic interaction, matching ability between host and guest. Thus, CD becomes one of the most important researching subjects of supermolecule chemistry [3].

With a large amount of chemically modified CDs being synthesized, the recognition ability and selectivity of CDs and their derivatives to the guest molecules increase constantly. It was shown that the stability of CD inclusions was directly influenced by factors such as size fitness between host and guest molecules, geometric complementation, etc and many weak interactions during the molecular recognition of CD. Through application of modern analysis methods [4], such as nuclear magnetic resonation, X-ray diffraction, infrared spectrum, Ultraviolet-Visible (UV-Vis) spectrum, fluorescence spectrum, electric chemistry and thermo analysis, supermolecule interaction mechanism and molecular recognition mechanism of CD can be easily revealed.

6.1.3. Research content and current status of CD supermolecule chemistry

Currently, in the research of CD supermolecule chemistry, besides constant synthesis of new CD derivatives and studying interactions between CDs and guest molecules, new progress has been achieved in the following aspects such as CD molecular recognition and self-assembly, rotaxane design and multi-rotaxane supermolecule, analogue enzyme, etc [5].

6.2. CD Analogue Enzyme

Enzyme is a kind of protein with special functions. It exhibits substrates and its reaction specificity could specifically catalyze some chemical reactions under mild conditions. Due to the unusual supermolecule interaction between enzyme and substrate, it is first used as the bionic micro-system [6].

Analogue enzyme (also called man-made enzyme or enzyme model) absorbed important factors in the enzyme catalysis, chemical, biochemical and molecule biological methods; designed some simpler protein molecules and non-protein molecules than native enzyme, and these molecules were utilized as model to analog enzyme catalysis. There are two main purposes to construct analogue enzyme: (1) to achieve high catalyzing efficiency, and enhance the enzyme tolerance of temperature, solvent and pH; (2) to help understand enzyme catalyzing mechanism, i.e. the source of high catalyzing efficiency of enzyme [7].

Thus, it is greatly significant to analog enzyme molecule recognition and catalysis function intelligently during the understanding of the biological evolution of enzyme itself and studying the structure and functions of the enzyme. It is one of the aims for scientists to design and synthesize the catalyst with efficient and stereoscopic specificity of enzyme or to illustrate the enzyme catalyzing principles completely [8].

6.2.1. Structure and catalyzing mechanism of native enzyme

6.2.1.1. Structure of native enzyme

Enzyme is a kind of highly efficient and specific biological catalyst. All biochemical reactions *in vivo* are enzyme-catalyzed reactions. High efficiency and high selectivity of enzyme originated from the hydrophobic interaction between enzyme and substrate and the adjacent effect of catalyzed groups [9].

Generally, enzymes with high activity are spherical proteins, of which multi-peptide chains are folded widely thus a dense structure is formed. Hydrophilic group of amino acids are distributed mostly outside the surface, but the hydrophobic ones are hidden within the inner space of the structure [10].

There are various kinds of functional groups in the enzyme molecule, such as $-NH_3$, $-COOH$, $-SH$, $-OH$ and so on. They have versatile functions and properties, among which some special functional groups were directly related to the enzyme catalytic activities, such as iminazole group of His, hydroxyl group of Ser, sulphuric group of Cys and carboxylic group of Glu and Asp. The active site of enzyme is composed of combining site and catalyzing site, the former combining with the substrate decides the specificity of the enzyme, while the latter acts as the catalyst [6].

Different enzymes have active sites constituted of different groups and various conformations. For simple enzyme (the composed unit of enzyme is only amino acid), the active site consists of a few near amino acid residue or side chain groups in these residues in the stereoscopic structure of enzyme molecules. They could be far from each other in the first structure and even in different peptide chains, but they are close to each other within the spacial conformation [7].

For combined enzyme (the enzyme needs auxiliary enzyme or auxiliary group), active site includes some chemical structure of auxiliary enzyme or auxiliary group besides amino acid group residues. Auxiliary groups are mostly inorganic ions or small organic molecules, of which they are likely to be composed partly of active site and act as bridge between the substrate and enzyme, or stabilize the catalyzed active and necessary molecule conformation of enzyme proteins, or transfer hydrogen, electron, some special atom in the catalyzing reaction, or act as the carrier of some unusual groups [7].

Generally, there is one active site of enzyme molecule, but there might be a few active sites of some enzymes composed of many sub-units. Generally, one catalyzed center includes side chains of 2–3 amino acid residues. Combining site is different with substrates of various enzymes and reaction properties; some have only one, while the others have several sites, and due to the different substrate properties, the number of amino acids included in combining center of various enzymes might be greatly different.

In the enzyme with a known structure, active site usually locates in the crack and crevice structure of the enzyme. The hydrophobic structure of crack and crevice is in favor of the combination between enzyme and substrate and excludes the entrance of water molecules. But when water molecules also participate in the reaction, crack and crevice need necessary radical residues to construct the required unusual environment for the catalytic reaction.

The specificity of combination between enzymes and substrates depend on the specific spacial scattering pattern of each related atoms of the active site. Substrate is required to have the matching shape which is adaptable to the structure of active site. Meanwhile, the shape of the active site could also change with combination of the substrate. A more complementary shape is formed with the substrate by induction action.

Noticeably, active site formation requires enzyme protein having a certain spacial conformation, but it is necessary that amino acid groups beyond the active site of enzyme molecule could stabilize the special conformation. Thus, it has great importance to the catalytic activity of enzyme, correspondingly. For instance, in these amino acid residue groups, some are related with the enzyme activity modulating, some are correlated to correct spacial structure formation of enzyme molecule and some are related with the immunogenicity or other characteristics of enzyme molecule.

6.2.1.2. *Catalyzing mechanism of native enzyme*

The catalyzing mechanism of the enzyme mainly includs approaching and orientating effect, conformation changing effect, acid-basic catalyzing mechanism, covalent catalyzing mechanism, micro-environmental effect, self-scissoring mechanism and self-splicing mechanism.

6.2.1.2.1. Approaching and orientating effect

Enzyme molecule has its active center. Groups on the active center could approach each other substrate, and make the substrate groups or catalyzing groups on the active center of enzyme be orientated according to correct position, thus favoring formation of the mid-product and sustaining the catalytic reaction.

Approaching and orientating effects between enzyme and substrate have the following roles for catalytic reactions:

(i) The concentration of substrate molecules near the active center could increase, thus accelerating the reaction velocity.
(ii) Groups of active center of enzyme might have track-orientation effect on the substrate molecules, thus decreasing the required active energy of reaction.
(iii) Enzyme reacts with substrate to produce mid-product ES, which could make the intermolecular reaction change into intramolecular reaction, thus increasing the reaction velocity. Meanwhile, the life of ES is about 10^{-7}–10^{-4} s, however, the average life of both molecules colliding randomly and integrating together is about 10^{-13} s. The former is 10^6–10^9 times the latter, which makes the probability of catalyzing reaction of enzyme increase significantly, resulting in higher reaction efficiency [11].

6.2.1.2.2. Conformation changing effect

When the enzyme molecule is close to the substrate molecule, due to the interactions between each other, both the enzyme molecule and substrate molecule could have conformation changes. Thus it is more beneficial to combine with substrate and react, and increase the reaction velocity greatly.

(i) Substrate induces the conformation change of enzyme molecule. Koshland put forward induced fit theory in 1958. He thought the conformation of enzyme molecule was not unchangeable, while as the substrate molecule appeared close to the enzyme molecules, the latter was induced by the former, thus the conformation of enzyme had some changes and made it in favor of combination with the substrate molecule.
(ii) Enzyme molecule induces the conformation change of substrate. When the enzyme molecule closes to the substrate, the substrate would have all sorts of twisted deformation or conformation changes under the induction of enzyme molecule in order to combine with active center of enzyme so well. Such a changed conformation is much like a transition state which makes the reaction active energy decrease greatly, and thus accelerate the reaction on going [11].

6.2.1.2.3. Acid-basic catalyzing mechanism

Acid–basic catalyzing mechanism considered the catalysis of enzyme as the proton transferring between enzyme and substrate, i.e. mutual transformation between acid and basic, thus decreased the required active energy of reaction and made the reaction accelerate. Acid and basic catalyzing groups of enzyme protein are provided by the side chain of amino acid residual groups, which is the most common with the iminazole group. Under the physiological condition, iminazole group is able to act as both proton donor and proton acceptor, and the speed of accept and release proton is very fast and equal.

Acid catalyzing group as a proton donor in enzyme protein is a kind of conjugate acid, while basic catalyzing group as a proton acceptor is a kind of conjugated basic. The catalysis of conjugated acid is firstly conjugated with the oxygen on the substrate carbonyl to generate the hydrogen bond to enable the carbon atom on the carbonyl bring much more positive charge, attract the unpaired electrons of water molecules more easily and decrease the active energy of covalent bond between carbonyl group and water molecules. Then, conjugated acid transfers H^+ into carbonyl oxygen of the substrate molecule and it becomes conjugated basic attracting H^+ of water molecule to form the hydrogen bond. This makes the oxygen of water molecule electro negativity increase, and conduct nucleophilic attack to carbon atom of carbonyl group more easily. This action decreases the active energy of carbon–oxygen bond generation and accelerates the reaction speed. The conjugated basic attracts H^+ of the water molecule to become conjugated acid, then transfers the attracted H^+ into one group of substrates and returns to the conjugated basic, thus the catalysis is completed [11].

6.2.1.2.4. Covalent catalytic mechanism

During the catalysis, the enzyme first combines with the substrate to form a kind of mid-complex, which is a covalent mid-product due to some group of enzymes attacking some special group of substrates. According to the different groups of enzymes attacking the substrates, covalent catalysis could be divided into nucleophilic catalysis and electrophilic catalysis.

Groups with abundant electrons (nucleophilic groups) of the enzyme molecule attack the groups with fewer electrons (electrophilic groups) of substrate molecule to form the covalent mid-product, which is called nucleophilic catalysis. The common nucleophilic catalysis has nucleophilic substitution and nucleophilic addition. Electrophilic catalysis is a procedure where groups with fewer electrons (electrophilic groups) of an enzyme molecule attack the abundant electron groups (nucleophilic groups) to form the covalent electrophilic mid-product [11].

6.2.1.2.5. Micro environmental effect

Micro environment is a kind of special hydrophobic reaction environment located by the catalytic group of the active center of enzyme. Due to this special hydrophobic reaction environment affecting the combination of enzyme and substrate, and influencing the dissociation of catalytic group, the reaction is accelerated, thus this kind of effect is called micro environmental effect.

What's more, enzyme catalyzed reaction by self-scissoring and self-splicing mechanism [11].

6.2.2. *Preparation mechanism of CD analogue enzyme*

Analogue enzyme is a kind of non-protein molecule which is similar to the function of native enzyme but its structure is much simpler. Analogue enzyme could be taken as a model to analog the combination of enzyme with substrate in order to increase reaction and stereoscopic selectivity. Research of analogue enzyme is an important content in the field of supermolecule chemistry [12].

CD has a unique cone-type molecular structure, its primary and secondary hydroxyl groups are located in the smaller and larger open end of cone respectively, which establishes the outside CD molecules hydrophobic. Its inner side is a cavity composed of hydrogen on C-3, C-5 and glycosidic oxygen and has hydrophobicity. So, CD could generate supermolecular inclusion with many guest molecules. It is not only an excellent mainbody in the molecular recognition field, but also a semi-natural acceptor in the analogue enzyme research. CD derivatives after selective modification are further improved in the aspect of structure (extension of hydrophobic zone) and properties (entrance of functional group). Thus, are specially fitted to become a mainbody material of constructive analogue enzyme [8].

To sum up, CDs have the following characteristics of an ideal analogue enzyme:

(1) An excellent water solubility, which could bond substrate selectively in the aqueous solution. Primary and secondary hydroxyl groups stereoscopically distributed on the two sides of the cavity could participate in the reaction with the substrate;
(2) Secondary hydroxyl groups have different proton donor activity, which could introduce catalyzing group and proton transferring system by chemical methods;
(3) Molecular cavity is composed of C_3-H, C_5-H of glucose residue and 1,4-glycosidic oxygen, which is a relatively rigid hydrophobic cavity, and is beneficial to the selective combination and positioning of organic molecules [5].

CD molecule itself is able to catalyze some chemical reactions due to the hydrophobic cavity similar to the native enzyme, for example, ester hydrolysis, alkylations of haloid, Diels–Alder reaction and so on. But the combination constant of CD and substrate is usually 10^4 mol/L, is less than that of enzyme and substrate. In addition, the catalyzing ability of hydroxyl group of CD is very limited, and is mainly concentrated on the modification of CD for constructing CD-based analogue enzyme, i.e. group with molecular recognition or catalyzing function could be introduced onto the surface of CD so as to improve the hydrophobic combination and catalyzing function of CD.

Generally speaking, the construction of a CD-based analogue enzyme needs to consider all kinds of characteristics of CD molecular structure to carry on, such as spacial matching between substrate and CD, association effect of catalyst, construction of microenvironment, selection of combining site and combining ways, metal ions auxiliary and construction of double combination site and so on. That is, the combination ability of the CD cavity should be effectively integrated with catalyzing groups and these special groups should be smartly matched in order to make them have a certain mutual relationship in the space. Commonly, the larger and more complicated the substrate molecule, the more beneficial it is to generate the energetically fitted combination site. Thus, multi-site combination is an effective design scheme of analogue enzyme.

Modification of CD's main surface and side surface is an important factor for construction enzyme model to consider. It was found effective to introduce catalyzing groups into main surface or side surface, but some substrates were much fond of the side surface (the front open is larger). Under such a condition when the catalyzing group was conjugated into the side surface, better activity was shown. If two functional groups were introduced into one side of the CD, both groups would produce synergy action and generate a new system which could improve the catalytic activity of CD. Meanwhile, due to both groups participating in the catalytic reaction with synergy, static electronic environment of bond-linkage plateau of CD could be kept within the required range. For example, when 1,4-dihydro nicotinamide (NAH) was introduced into the main surface of the CD, the velocity of reducing hydro ninhydrin was 50-fold that of reduced form of nicotinamide-adenine dinucleotid (NADH). When two NAHs were simultaneously introduced into main surface of CD, the obtained analogue enzyme could increase its catalyzing efficiency 100-fold for reducing hydro ninhydrin.

Two iminazole groups were simultaneously introduced into main surface of CD, which could analog ribonuclease A (RNase A) [13]. K_{cat} of such an analogue enzyme catalyzing cyclic annular phosphate (CAP) is 120×10^{-5} S^{-1}, which is much higher than that of non-enzymatic reaction ($K_{uncat} 1 \times 10^{-5}$ S^{-1}) [14]. Such an analogue enzyme shows an excellent stereoselectivity where the ratio of two

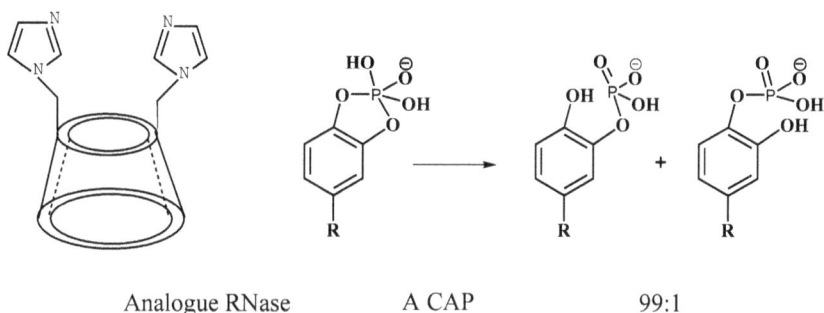

Analogue RNase   A CAP   99:1

Fig. 6.1. CD Analogue RNase A catalyzed to hydrolyze CAP.

kinds of phosphate ester by such enzyme catalysis was 99:1, while non-enzymatic catalytic products of CAP in the NaOH solution was 1:1 mixture (Fig. 6.1).

It was indicated by isotope and dynamics studies that two iminazole groups play a synergetic role in catalysis, which acted as acidic catalyst and basic catalyst respectively. In addition, it was much important for catalysis that the relative positions of two iminazole groups on β-CD exist. As long as two iminazole groups were located at adjacent positions, analogue enzyme could have higher catalytic efficiency and stereoscopic selectivity [15].

Groups with special conformation were introduced into CD. This could get analogue enzyme model with stereoscopic selectivity. For example, pyridoxamine (PM) was bonded to the first hydroxyl group of β-CD. The obtained modified CD could catalyze amino transfer reaction of α-ketone acid. The reaction speed was 100 fold that of PM as catalyst and the stereoselectivity was shown during the catalysis [16]. If the PM group was linked to the main surface of β-CD through double bonds, the transaminase model would be obtained. It was indicated from studies that when PM group was located on one side of CD, the substrate of meta substituted phenyl pyruvic acid would be linked as priority selection; when it was located above the cavity of CD, the substrate of p-substituted phenyl pyruvic acid would be linked as priority selection [17].

In addition, after the CD units were bridged by some functional groups, two adjacent CDs' cavities were able to synergistically participate in the inclusion and complexation of guest molecules with the suited shape and size. More stable supramolecular complexes would be formed to better analog the biological enzyme. Based on this, a series of CD polymers have been synthesized by researchers [18].

Metal ions could play an important role in many electro transferring systems, which was frequently used as super acid catalyst and had directional function. Thus, metal ions were introduced into analogue enzyme model in order to increase its catalyzing efficiency. The earliest CD and metal complex was carboxypeptidase model. One or two oxime-modified CD were complexed with Cu(II) or Ni(II) ions, which could catalyze the hydrolysis of substrate complexed in the cavity of CD by metal ions [19].

CD analogue enzyme could be also used in the bionic sensor. For example, the NAH group was introduced to the side surface of β-CD to analog dehydrogenase and used into bionic sensor. The inclusion ability of CD and electron transferring ability of NAH group to alcohol, acetone, dopamine and propanol were observed by electro-chemistry and fluorescence technology under the conditions mixed with hydrophilic and hydrophobic negative ions. It was found that the size and hydrophobicity of the substrate could directly affect the catalytic activity of the enzyme model. Such a man-made sensor could be utilized repeatedly.

6.2.3. *Typical CD analogue enzyme models*

Currently, great advances have been achieved in the studies of CD analogue enzyme. Many biochemical reactions have been simulated intelligently, among which the most studied analogue enzyme is the metal hydrolysis enzyme. The earlier enzyme model (Fig. 6.2(a)) is an analogue hydrolysis enzyme for *p*-nitrophenyl acetate (PNPA) reported in 1970 [21]. Imidazole group was introduced into primary or secondary hydroxyl group of CD of such a model. When an imidazole group was directly bonded with primary hydroxyl group of β-CD, only small catalytic activity was shown, but when the secondary hydroxyl group of β-CD was correspondingly modified, the catalytic activity was 1,000-fold that of the unmodified one at pH 7.5. The ratio of catalytic activity of such two kinds of modified CD is more than 70, which indicates an obvious position effect. The big difference in catalytic activity is due to the different direction between active groups and reactive center (carbonyl carbon in substrate). The nitryl taken as the head of substrate might enter into the

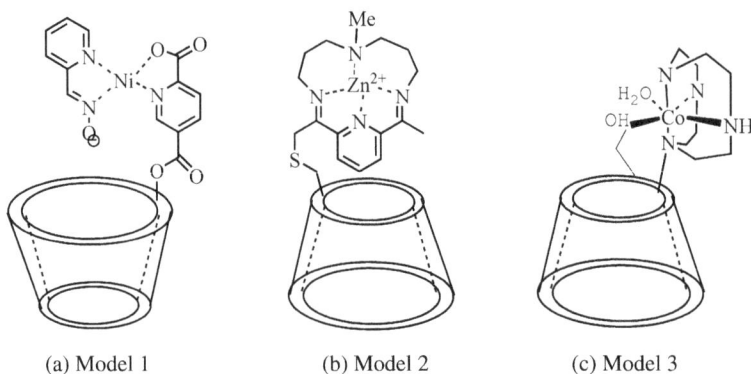

(a) Model 1   (b) Model 2   (c) Model 3

Fig. 6.2. Metal hydrolyzing enzyme models based on CDs.

cavity from the side of secondary hydroxyl group of CD, which is more suitable for secondary hydroxyl group modified CD [20].

When two imidazole groups were introduced into one side (the first side) of primary hydroxyl group of β-CD, due to synergetic effect of both groups, charge system for renewal was formed, thus selective catalytic hydrolysis effect was achieved. Two imidazole groups in the model could have three kinds of bonding ways: AB, AC and AD (according to glucose order). Only AB isomer had the best hydrolysis effect for the substrate of phosphate ester butyl phenol at pH 7.0, which indicated the high stereoselective hydrolysis effect. The geometric construction of AB isomer combined with the model substrate was analyzed. Imidazole group could be thought of as the acidic catalyst where the phosphate anion protons made the P-O-1 of the substrate preferential fracture and water molecules were transferred by neutral imidazole group. It was shown that mid product could be directly transferred into the product and no rotation was required, meanwhile, catalytic group did not change and could react continuously, thus increased the turnover number of the reaction. The reaction constant of such an analogue enzyme catalyzing CAP hydrolysis is not very high, but it posesses a typical kind of enzyme activity.

After that, the built-up Zinc-enzyme model (Fig. 6.2(b)) could effectively catalyze the hydrolysis of aryl phosphate. Another better hydrolyzing enzyme model is 4 nitrogen heterocyclic 12 silane cobalt complexes (Fig. 6.2(c)), which is one of the most effective acyl transferring catalysts. If the 4 nitrogen heterocyclic 12 silane cobalt complexes were covalently linked with β-CD, the obtained enzyme model could increase 900-fold of the hydrolysis rate of PNPA (*p*-nitrophenol acetate). If the same group was conjugated onto the second hydroxyl group of β-CD, the obtained model could increase 2,900–3,700-fold of the hydrolysis rate of PNPA [21].

One analog enzyme called β-benzyme was synthesized in 1985, the catalyzing activity of which has the same order of magnitude as the natural chymotrypsin, and the stability of which is high. Such a model utilized CD as the substrate binding site, the carboxyl group of the modified group linked on the second carbon, the imidazolyl group and one second hydroxyl group of CD all together composed of the similar catalyzing active site of natural enzyme and thus the analog holoenzyme of chymotrypsin was realized (Fig. 6.3). The hydrolysis rate of tertiary phenyl butyl acetate by β-benzyme is above one-fold than that by natural enzyme, K_{cat}/K_m value of β-benzyme is also equal to that of natural enzyme [20].

Transaminase model is another example of synergistic interaction by introducing two active groups. Usually, natural transaminase was firstly required to activate zymogen by coenzyme phosphopyridoxamine (PMP, Vitamin B_6) in order

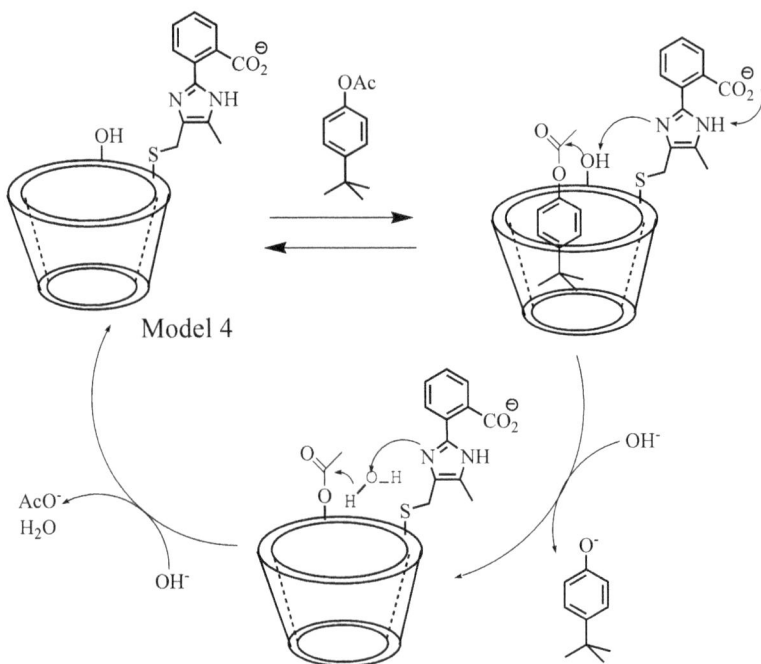

Fig. 6.3. Analog chymotrypsin (β-benzyme) model and its catalyzing mechanism.

to catalyze various amino acid reactions, such as decarboxylation, elimination, transamino reaction and so on. The PM group bonded to the first hydroxyl group of CD is able to form an analog enzyme model (Fig. 6.4) which could simulate holoenzyme.

Transaminase model I could make the reaction rate of indole pyruvic acid transferring into indole alanine being 200-fold faster than PM, which makes the enantiomer surplus percentage of L-Trp generated from indole pyruvic acid and L-Ala generated from phenylpyruvic acid reach up to 12% and 67%, respectively. The reaction had obvious optical induction [17].

It was indicated from studies that at the active site of natural enzyme dependent on Vitamin B_6, aldehyde group and ε-amino group of lysine residue could form Schiff base. Under various noncovalent interactions coenzyme is embedded in the enzyme protein. There are three bonds to be dissociated effectively by electron dipping effect of conjugate 4-pyridinyl around α-carbon in the part of amino acid of intermediates at the transition state [22]. On the basis of above, a new transaminase model structure was further developed. Taking AB type sulfonic acid ester generated from *m*-benzenedisulfochloride and β-CD as intermediate,

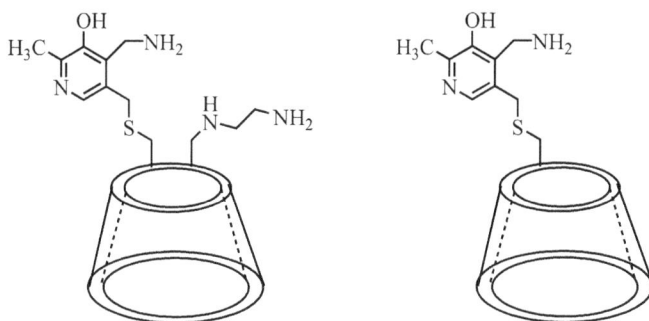

Transaminase model I   Transaminase model II

Fig. 6.4.   Analog holoenzyme model of transaminase reaction.

Table 6.1.   Optical activity of α-ketonic acid translated into α-amino acids.

α-ketonic acid	α-amino acids	L/D
R = Ph—CH$_2$—C=O-COOH	Ph—CH$_2$CH(NH$_2$)COOH (Phe)	98/2
CH$_3$—C=O-COOH	CH3—CH(NH$_2$)—COOH (Ala)	98/2
		95/2

after transferring into diiodo-CD, transaminase model II could be obtained by step-by-step reaction with PM and ethanediamine. Such a model could make the transamino reaction of ketonic acid have a high chiral selectivity and the generated L-Trp could reach up to 95%, L/D ratio of catalyzed phenyl L-amino acids is listed in Table 6.1. Plate intermediate C…N…C is formed in the reaction. Ethylenediamine group as the proton transferring catalyst is kept stationary on one side of the plate, which makes the reaction have a high enantiomer selectivity. From above, if the microenvironment provided by enzyme protein was not considered, the obvious increasing reaction rate, high substrate selectivity and reactive selectivity could not have been reached [23].

Recently, a series of CD ox-redo enzyme models have been prepared. For example, analog amino acid oxidase is built by the complex generated from CD-bi-*m*-carboxyl benzene sulfonic acid ester and ferric trichloride (Fig. 6.5).

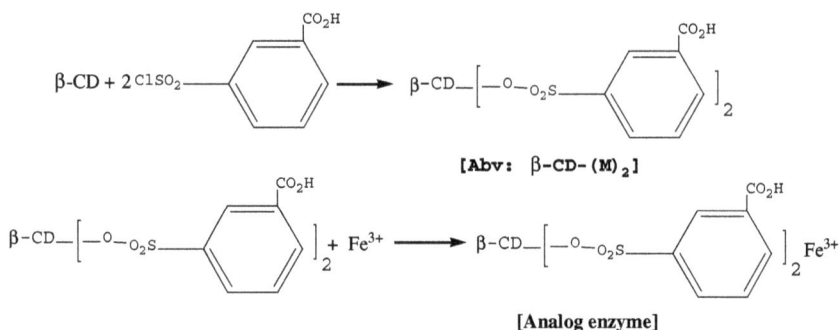

Fig. 6.5.   Preparation reaction of analog amino acid oxidative enzyme.

The phenylalanine is oxidized and deaminized to generate phenylpyruvic acid, under the presence of surplus hydrogen peroxide. Phenylpyruvic acid is further oxidized and decarboxylated to generate phenylacetic acid. If such a reaction system is under the condition of weak base environment of dimethyl formamide (DMF), the phenylalanine will be oxidized into tyrosine [24].

Analog glucose oxidase could be constructed by the complex formed from β-CD-di-*m*-carboxyl benzene sulfonic acid ester and ferrous chloride (Fig. 6.5). Enzyme-substrate complex could be formed by such an analog enzyme and hydrogen peroxide which could specially catalyze glucose to be oxidized into gluconic acid. The catalyzing dynamics of such an analog enzyme could be characterized by typical absorbance at 365 nm of the product generated from gluconic acid and toluidine, catalyzing efficiency of which is much higher than that of inorganic catalyst ($FeCl_3+H_2O_2$) [24].

Furoic acid could be prepared by furfural oxidation. Furfural is generated by pentose dehydration in the condition of 3%~5% dilute sulphuric acid, whereas pentose is included in the agricultural by-products such as peanut hull, corncob, wheat straw, boll hull, bagasse and so on. Furoic acid is an important industrial raw material to synthesize resin, medicine, pesticide, etc, and has a vital economic value. There are many methods of furoic acid preparation from furfural oxidation, such as cannizarro's reaction, permanganate method, potassium chromate method, sodium-hydrochloride process, hydrogen peroxide method, pure oxygen method and so on. In these chemical methods, some could produce a lot of by-products; some could use toxic or expensive chemical catalysts. The new catalyst was built by β-CD. Furfural could be oxidized into furoic acid within 1 h at the condition of acidic environment with 50°C, the catalyzing efficiency of which is 2.7×10^4 fold that of H_2O_2, 64-fold of $FeCl_3 + H_2O_2$, the transferring ratio could reach 97%. Such a catalyst could precipitate in organic solvent such as acetone,

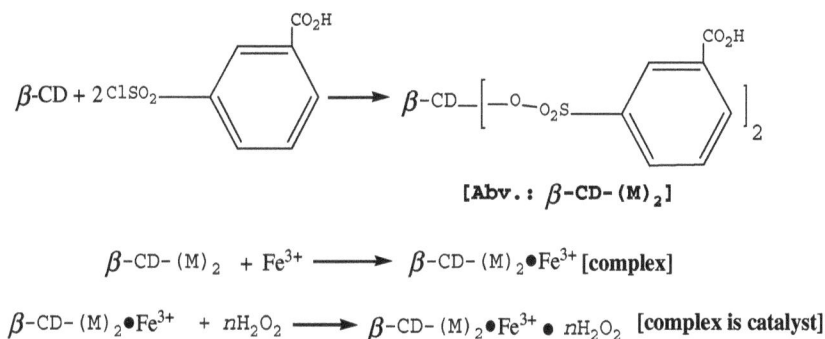

Fig. 6.6. Preparation reaction of analog CD complex oxidase enzyme.

thus could be separated from furoic acid, and could be used repeatedly. It has the following advantages: easy circulation and regeneration, low cost, nonhazardous, convenient, high catalyzing efficiency and short reaction time [25].

Preparation reaction of analog CD complex oxidase enzyme is schemed in Fig. 6.6.

In addition, a kind of CD analog enzyme has been prepared which could selectively oxidize the substrate with different spacial confirmation. Currently, CD analog enzyme could be used not only to catalyze chemical reaction, but also to supply the probable spacial confirmation of transition state.

In recent years, great progress in analog enzyme research has been achieved by using the bridging-CD. Catalyzing active center with synergistic inclusion and multiple recognition function consisted of two CDs and the functional group of the bridge groups. Bridging-CD could better simulate recognition and catalyzing function of analog enzyme to substrate. For example, triethanolamine bridged-CD dimer could hydrolyze di (*p*-nitrophenyl) carbonic esters, what's more, the catalyzing efficiency of which increased 150 fold that of control reaction [26].

Breslow has done a lot of studies on bridging-CDs in recent years, which not only reported a series of synthetic methods and binding capacity of the bridging-CD, but also successfully applied an analog enzyme into catalyzing a hydrolyzing reaction of double hydrophobic organic ester. When the substrate (the organic ester with double hydrophobic positions) was entrapped by cavities of double CDs, Cu^{2+} coordinated with the bridge located near the substrate ester group, which is beneficial for $-OH$ to attack the ester group, thus could remarkably accelerate the hydrolysis reaction (Fig. 6.7). The catalyzing rate increased 1.8×10^4 fold that of non-enzymatic catalytic reaction [8–10, 12, 13, 15, 18, 27–35].

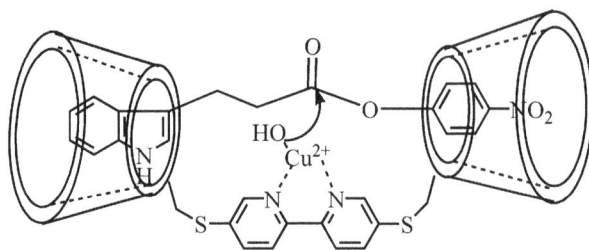

Fig. 6.7.   Hydrolyzing mechanism for ester under the copper ions by bridged CDs.

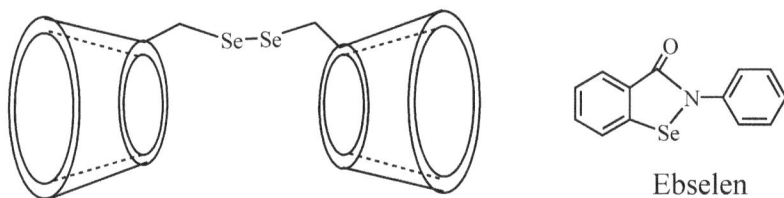

Ebselen

Fig. 6.8.   Analog GPx.

Similarly, analog glutathione peroxidase (GPx) could be constructed by 6-Se bridging β-CD. GPx could take reduced glutathione as substrate, and have an obvious effect on the treatment and prevention of cardiovascular disease and cancer by catalyzing reduced hydroperoxides, removing the internal free radicals and inhibiting lipid peroxidation. Bridging-CD was synthesized by using the typical structure of β-CD and introducing active group Se of GPx into the bridge chain in the studies. In Comparing the bioactivity of Se-bridging β-CD analog enzyme to other small molecular analog enzyme (such as ebselen (2-phenyl-1,2-benzene isoselenium pyrrole-3(2H)-one)) (Fig. 6.8), it was found that Se-bridging β-CD had a higher bioactivity than GPx, which might be caused by the increment in the cavities of CD as substrate binding sites [36].

For example, analog enzyme such as 2-Se or 6-Se bridging β-CD could show an excellent GPx activity, the bioactivity of which is above 4.3-fold of famous analog GPx ebselen [37]. It was also found that the substitution position on 2- and 6- of bridging β-CD might have a remarkable influence on its activity, and adding catalyzing group could increase the reaction activity. When tellurium (Te) was introduced onto 2- and 6- positions of the β-CD, it was found that these new analog CD models showed a higher GPx activity, among which the activity of 2-Te bridging β-CD might be 47-fold that of Ebselen [38].

Besides hydrolyzing enzyme [30], oxidoreductase [39], RNase [13], transaminase [40], CD could be used to build up many analog enzymes, such as carbonic anhydrase, thiaminase [41], hydroxyl aldehyde condensation enzyme [42], biotin [29] and so on, and all of them had an excellent result.

6.3. Molecular Recognition and Self-Assembling

Molecular recognition and self-assembling is a common phenomenon for maintaining life. Natural supermacromolecules systems constructed by molecular recognition and self-assembling exhibit various multifunctions, for instance, phospholipid, saccharide esters, glycoprotein and transmembrane spiral peptide etc. With employment in this way, they consist of ordering and mobility macromolecules for switching energy, converting messages, and delivering substances during a life process. However, how do these supermacromolecules form *in vivo*? Currently, more and more scientists and researchers are interested in how to imitate their functions and further reveal the significance of underlying mechanism in the design and construction of supermacromolecules to constitute novel molecular and electron devices. The high selectivity provided by molecular recognition is capable of cleaning tons of hurdles in life science, materials science and separation science.

Generally, molecular recognition and self-assembling of supermacromolecules exhibit the following features:

(1) Economical and highly centralized;
(2) Effective rapid generation of various self-assembling units with stable structures from simple subunits;
(3) Achieve the economical minimum of the amounts of structural information in self-assembling units by the interactions between subunits;
(4) Molecular recognition based on the interactions between weak noncovalent force and molecular form dynamic reversible self-assembling units with minimal energy in thermodynamics.

Compared with the molecular maintenance by covalent forces in traditional molecular chemistry, supermacromolecules obtained by controlling the bonds among molecules can make use of noncovalent forces to form self-assembling units of different atoms and molecules. The novel materials were developed by classical chemistry and supermacromolecular systems, and then assembled into the device, especially nanomaterials and their compositional device, which has a great significance in sustainable development for information technology, life science, novel material and ecosystem.

6.3.1. *Molecular recognition of CD and its derivatives*

Molecular recognition can be considered as processes where the subject selectively combines with object in generating certain functions. It is the basis for the function of assembling units and the core of supermacromolecules. The molecular recognition of supermacromolecules and chirality characteristics has key theoretic and practical significance for imitating biological function at molecular level, in the study of converting messages and in the process of binding of enzymes with substrates. Furthermore, the applications of molecular recognition offer broad prospects for analytical methods with high specificity and selectivity. The accurate recognition is generally based on the interaction between subject and object, including van der Waals force, electrostatic attraction, hydrogen bond and hydrophobic interactions. Molecular recognition and assembling by synergistic reaction of weak interaction is a common phenomenon in nature. When the shapes and binding sites of subject and object are complementary in geometry and electronic shell, a strong molecular and chirality recognition occurs [27].

6.3.1.1. *The molecular recognition mechanism of CD*

Due to the hydrophobic cavity and hydrophilic surface in the CD, it can be utilized as a subject compound to form encapsulation combining with inorganic, organic and biological molecule. This encapsulation depends on size-matching, geometry complement, hydrophobic interactions, van der Waals force, interactions of hydrogen bond or dipole–dipole, and some weak interactions for instance, releasing high-energy water and alleviating ring strain. CD encapsulation is highly correlated with the polarity, ionized state and chirality spatial structure. For example, a molecule with less polarity is more intended to encapsulate with hydrophobic cavity in the CD. Objects with non-ionized state are more active with the CD than those with ionized state. This selectivity exhibits the molecular recognition mechanism of CD for objects. Generally, the cavity size of α-CD is suitable for the encapsulation of monocyclic aromatics (benzene, phenol, etc). In contrast, β-CD is more suitable for naphthalene ring. Moreover, γ-CD gives preference to triaromatic hydrocarbon, for instance, anthracene and phenanthrene, for forming stable structure encapsulations. γ-CD can generate ideal crystal structures with 12-crown-4 by encapsulation. The CD is capable of selecting different substituent groups as objects for encapsulation. For instance, during the process of the encapsulation of β-CD with naphthalenesulfonate, β-CD is adjustable to the selectivity of objects in the subject and can recognize the slight difference of objects due to the synergistic reaction of weak interactions. Therefore,

β-CD is sensitive to β-naphthalenesulfonate and α-naphthalenesulfonate with 100% selectivity [43].

6.3.1.2. *Enhancement of the molecular recognition of CD by chemical modification*

Due to the limitation of the molecular recognition capacity of the natural CD, the design and synthesis of novel CD derivatives with functional groups and exploring the rules of the interaction recognition between object–subject molecules and the properties of functional groups are fundamentally significant for the study of supermacromolecules. The chemical modification of CD can be employed to modify the three-dimensional structures, enlarge the binding of cavity and provide the specific geometry shape for specifically fitting the subject or producing the special chirality site. For instance, the natural CD is composed of D-pyranose displaying excellent encapsulation selectivity for chiral enantiomer. By the electrostatic interaction between copper ion and amino acid containing negative anions, the complex of CD-copper improves the binding capability for amino acid. Concurrently, the formation of this complex changes the original molecular configuration of CD, exhibiting the high selectivity of the encapsulation of D-amino acid.

The introduction of chromophore into CD activates the weak interaction between hydrophobic cavity of CD and modified arms. When encapsulating the molecule of object, the binding capacity for the molecule of object is allowed to be enhanced through $\pi-\pi$ acting force between substituent group and the molecule of object and π-CH hyperconjugation. Moreover, the introduction of aromatic nucleus chromophore can alter some physical characteristics including light, electricity, magnetic force and energy conveying. For instance, the encapsulation of artificially synthesized substitute of CD derivative with merocyanine leads to the transformation of absorbed energy from light into merocyanine efficiently [24].

In addition, chromophore generated from CD can be also used as the probe of spectroscopy, further enabling fluorescence, UV and circular dichroism spectra in the application of the molecular recognition. For instance, the electron deficiency core in the phosphate of CD provides a recognition site for the multi-recognition of amino acids. The functional groups are seen as the molecular probe in UV, for recognizing the size and chirality of the amino acids. Another case is in the followings the introduction of transition metal into β-CD derivative can improve the capacity of the recognition of chemicals by encapsulation and collocation, especially for adamantane derivative (330 fold increase). During the study of the recognition of amino acid CD for the molecule of fatty alcohol and its chirality, the encapsulation capacity of C10 alcohols was found to highly correlate with the size, shape and rigidity of object. For instance, adamantanol is highly selective for nerol (70%). However, for some farnesols in C10 alcohols with similar structures the selectivity of 2-adamantane alcohol for menthol is upto 32% [30, 31].

It is well-recognized for many years that the synthesis of bridging-CDs as an effective supermacromolecule and its molecular recognition requires much attention. Bridging-CDs composed of two hydrophobic cavities can combine with metal ions leading to its linkage and clathration properties. Utilizing synergistic encapsulation reaction and multi-sites recognition mechanism is able to boost the ability of the specifically binding subject of the bond and the selectivity for molecules. For example, for Se–platinum modified CD system, some of platinums are effective antitumor agents; Se can be used as HIV-1 transcription inhibitors. Both *o*-phenyldiselenium bridging β-CD and the complex between *o*-phenyldiselenium bridging β-CD and platinum were taken as host molecules to coordinate with 8-phenylamino-1-naphthalenesulfonate (ANS) and produce molecular inclusion. It was found that the bonding constant between *o*-phenyldiselenium bridging β-CD and ANS is 12.4-fold that between natural CD and ANS, whereas the bonding constant between the complex between *o*-phenyldiselenium bridging β-CD and platinum and ANS is 3-fold that between *o*-phenyldiselenium bridging β-CD and ANS [44].

As mentioned above, the change in the structure of CD can control the capabilities of molecular linkage and clathration, showing the significance of the lock–key theory for selectively binding the substrate, further demonstrating the importance of multi-recognition mechanism in the encapsulation of CD [45].

At present, the molecular recognition of CD and its derivatives has been successfully applied in antimer compounds, stereoselective separation in chromatography, modeling enzyme and enantioselective catalyzed reactions [46].

6.3.1.3. *The synergistic reaction of synergistic reaction CD with the subject recognition molecular — calixarene and crown ether*

Different objects, for instance, CD, crown ether, calixarene exhibit various properties and disadvantages for molecular recognition. If molecular recognition process could be coordinated, the selective recognition function can be improved effectively. For example, for gas chromatography (GC), the solid phase of the mixture including CD and crown ether or the mixture of CD and calixarene illustrates synergistic recognition in the process of the separation of aromatic isomers, allowing to the improvement of the selectivity for isomers. When introducing single crown ether into β-CD, the association constant is significantly higher than that of the control. For the synthesis of the chiral solid phase of 6-(1- benzoazepine-15-crown-5) -2, 3, 6-O-methyl-β-CD, the synergistic reaction of CD and crown ether ring increases the recognition sites for objects, enabling the enhancement of the interaction force between new synthesized solid phase and antimer as well as isomer to obtain a better performance of separation. For the synthesis of hydrophilic calixarene-β-CD fluoresced acceptor, the unit of CD and calixarene interacts with objects, for instance, atropine and menthol, making the fluoresced functional group exposed to water, further increasing the chance of fluorescent quenching upon the detectable level.

6.3.2. The self-assembling and assembling CD and its derivative

Diverse supramolecular assembling with special functions in nature could generally provide a highly ordered microenvironment. Enzyme and biological membranes are typical examples as survivors after long-term evolution. Artificial supramolecular assembling has several unparalleled advantages, for instance, the ease of synthesis, highly stable and universal molecular structures etc. So far, molecular assembling based on molecular recognition includes template effect, self-assembling and self-organization. The generation of supramolecular functional systems by molecular assembling is one of the goals of supramolecular chemistry. As the main molecule with tubular structures, CD can encapsulate with linear macromolecules, producing various supramolecular assemblings including catenena, rotaxane, polyrotaxane and nanotubes, which indicates the extensive potential in chemical engineering and material science [47].

6.3.2.1. Catenena and rotaxane

Catenena with a within-district cycle shape is the internal lock supramolecular composition consisting of two or more cycles connected by noncovalent bonds. The cycle structure including CD is one of the significant compounds for the manufacture of supramolecular composition, for example, catenena. This kind of complex assembly is helpful for scientists to understand the mechanism of noncovalent bonds when CD is applied in the linkage substrate. In 1993, Stoddart group successfully synthesized two types of catenenas. 2,6 methylated β-CD (DM-β-CD) can be synthesized to quasi-catenena with one hydrophobic aromatic cycle and two hydrophilic long-chain owning polyether side chain and the amido end. In the NaOH solution, dichlorobenzoyl chloride can be used to end the amido end for manufacturing several catenenas simultaneously [25].

Pseudorotaxane is produced when the linear molecule of the object penetrates the cyclic subject. While functional groups or compositions are attaching both ends of the object through the covalent bond or coordinate bond, the formation of a "stopper" shape blocks the separation of pseudorotaxane from the main body, generating rotaxane. Rotaxane and pseudorotaxane are both supramolecules maintained by the weak interaction of noncovalent bonds. The unit molecule determines the properties of the whole large molecule. The rotaxane is composed of a linear molecule and a cyclic molecule. N-rotaxane is formed when a linear molecule passes through n-1 cyclic molecules. Due to the special noncovalent supramolecular structure, this sort of supramolecule demonstrates the special character and has potential for application [48].

Rotaxane is an excellent example with the significance for studying and manufacturing this sort of supramolecule.

6.3.2.1.1. The synthesis of the molecule with special structure by rotaxane

During the process of synthesis of the cyclic molecule, the yield can be decreased when the linear polymerization for each monomer occurs. However, the yield can be improved by means of the introduction of objects into reaction, further forming rotaxane with the cyclic molecule, resulting in the cyclic synthesis predominating in the reaction.

6.3.2.1.2. Molecular switch and molecular shuttle

In particular, by the employment of functional fragments and recognition sites in rotaxane, the situation changes or mechanical motion occurs inside, leading to two different states under the stimulation of light and electrical signal or the change of the systemically chemical concentration detected by chemical method and spectroscopy. When the external signal disappears, rotaxane returns to the original condition. In addition, the cyclic molecule in rotaxane rotates or moves in parallel along the linear molecule [14].

CD can react with the linear polymers including some hydrophilic chemicals, for instance, polyethylene glycol (PEG), polypropylene glycol (PPG), and perfluoromethyl vinyl ether (PMVE), or some lipophilic chemicals, for example, low-density polyethylene (LDPE), polypropylene (PP), polyisobutylene imide (PIBI), polyester for the production of rotaxane and n-rotaxane. PEG is mixed with CD to achieve the precipitation. The yield increases with the increase in

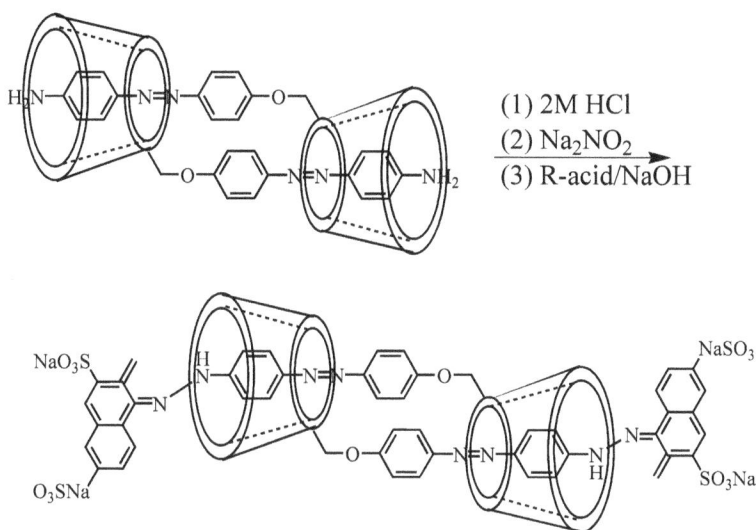

Fig. 6.9. Structural schematic illustration of catenane and its synthesizing methods.

polymerization. The results of ^{1}H-NMR indicate the ratio of PEG and CD is two when the degree of polymerization is more than six. Since the length of two units of glycol is corresponding to the cavity of α-CD, each of α-CDs in these polymers is contiguous to each other, as proved by the powder crystallization figure and cross polarization/magic angle spinning nuclear magnetic resonance (CP/MAS NMR). The hydrogen bonds interacting with α-CDs result in the stability of piped structure. The arrangement of α-CD is highly ordered. In addition, the generation of α-CD/PEG is reversible. When the employment of amounts of small molecule is encapsulated by α-CD, for instance, benzoic acid, the structure decomposes.

Analogous polymer can be synthesized including different units and CD with different stoichiometry ratios (Fig. 6.9). Some of them are hydrophobic. It is noted that the size of object is matched with the size of the cavity of CD.

In Japan, Kaneda with his coworkers synthesized quasi-rotaxane called Janus. First, azobenzene or quasi-substituted group was introduced into the 6-hydroxyl position of CD. Two CDs formed one group in the solution. Each substitutent group penetrated the inside of CD generating quasi-rotaxane, further synthesizing rotaxane [49].

The drive force of rotaxane through encapsulation of CD and polymers is derived from the hydrogen bonds of contiguous CD molecules and the hydrophobic and dimension interaction between the subject and the object. Therefore, some environmental factors, for example, temperature and dielectric have a direct effect on the state of rotaxane resulting from CD assembling in the polymer chain. Further, the molecular switch with special functions can be produced in terms of this theory. In 1997, Nakashima synthesized a molecular switch containing CD with photoactivity [50]. With the condition of photo induction, trans-azo segments penetrating the cavity of α-CD isomerizes cis-configuration decreasing the spatial matching, further leading to the separation of CD from azo segments and moving into methylene fragments of the linear molecule. During the isomerization of cis-configuration into trans-configuration, CD is relocated at the azo functional group indicating the switch process on the molecular level [50].

As for the blocking group, Phe could be connected with peptide bonds and PEG for manufacturing biodegradable polyrotaxanes formed by CDs. These polyrotaxanes with immobilized drugs can slowly release drugs by the disintegration of the structure of supermolecule. The affinity of enzyme to the peptide of the end group in PEG is not affected by this structure. Polyrotaxanes include fluoresce in isothiocyanate (the blocking functional group), β-CD and PEG–PPG–PEG. Induced circular dichroism and NMR results demonstrate most of β-CD shift from PEG to PPG with the increase of temperature. Hence, the motion of CD on the polymer chain can be controlled by the temperature, seemed as a molecular shuttle. Polyrotaxane includes α-CD and PEG with the end group taking place by azobenzene. With the treatment of the light, cis–trans isomerization of azobenzene occurs. Once cis–trans isomerization of azobenzene completes, the structure of polyrotaxane is stable. Therefore, azobenzene at the end group exhibits the properties of photo-timing gate to adjust the structural change of polyrotaxane.

A CD modified by 6-naphthalene monosulfonic acid and chain polyethers form rotaxane, further generating polyrotaxane where compositional cycles can move cyclically or linearly along with the main frame. This property sets it apart from traditional copolymers or quasi-mixtures. It allows this material to apply in the improvement of binding ability, handling characteristics, thermal properties, the control of viscosity and molecular arrangement.

6.3.2.2. *The nanostructure of metal complex of self-assembling*

The length of a nanoparticle is equal to 10 atoms. Nanotechnology specializes in the study a single atom or a small amount of atoms and molecular assembling at the micro level for producing the functional substances. Given the high correlation with a single atom and the control technology of molecule, this effective control on the micro level can be utilized to understand the structural mechanism pertaining to the properties of nanosubstances. With the development of nanotechnology, nanophysics, nanochemistry, nanomaterial, nanoelectronics, nanobiology have been extensively developed.

Through mercaptanon self-assembling on the surface of metal driven by the accurate balance of dispersion force in the long chain, it forms multifunctional single and multi-layers and inserts objects into pre-build subjects for the most effective method for the nanomixtures. In addition, taking advantage of this method can manufacture the ion and electric conductors. The combined assembling of organic surfactants and inorganic compounds generates a series of intermediary porous materials: Utilization of pre-build subjects such as rotaxane constructs nanomolecules complex with the exact construction and special function single-layer LB film, rotaxane and pipe by self-assembling.

CD and polyrotaxane are used as intermediates in the production of nanopipe CD complex indicating high selectivity for different macromolecules. For instance, CD reacts with PEG and both amino-group ends to synthesize quasi-polyrotaxane. The whole procedure can be described thus: CD attaches the chain of PEG, then 2, 4-dinitrofluorobenzene reacts with the amino-group of the chain of PEG blocking the formation of CD polyrotaxane separating from the chain of PEG. Subsequently, adjacent CDs on the chain of PEG are crosslinked and 2, 4-dinitrofluorobenzene as well as the chain of PEG in cyclodextrin are removed further achieving the corresponding nanotube CD complex. Moreover, α-CD polyrotaxane based on PEG reacts with epichlorohydrinin NaOH (10% w/w) obtaining nanotube CD complex after removing end capping reagent in NaOH (25% w/w). The final product exhibits a huge potential for the drug delivery.

Furthermore, β- and γ-CDs are able to form nanotubes with all trans-1, 6-diphenyl-1, 3, 5-hexatriene (Fig. 6.10) containing 20 β-CDs with double structure where two β-CDs are connected by every object when passing one β-CD. α-CD with smaller cavity fails to form the nanotube, indicating that the degree of congruence for different cavities of α-CD to hexatriene rod plays a significant role in the generation of nanotube. This kind of nanotube is also utilized in molecular wires or/and photo-switch with huge potential [51].

Crypt, CD and crown ethers reveal the specific selectivity on alkali metal ion enabling penetration into the thin film and combining with the sensitive site of field effect tubes, further developed for the alkali metal ion appliance.

6.3.2.3. *Artificial membrane*

Self-assembling membrane (SAM) is a new type of organic ultrathin membrane emerging in recent years. It consists of specific organic molecules synthesized by

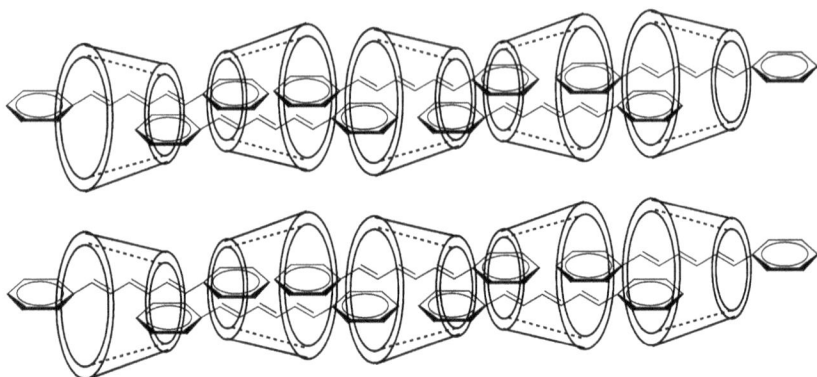

Fig. 6.10. Nanotube formed by self-assembling from β- and γ-CDs with all trans-1,6-diphenyl-1,3,5-hexatriene.

chemical bonds on the suitable solid material, producing the single molecular film in a tight ordered sequence. SAM is widely employed in biomimetic and study of physicochemical properties of layers.

CD derivative could be used in supermolecule building block and in the manufacture of various single molecules or multimolecular films. CD takes advantages of its characteristics combining with numerous organic compounds to generate subject–object assemblages. Additionally, CD derivative can be orderly assembled on the surface of solid electrodes, allowing the imitation of transfer process of biomembrane. This finding plays an important role in molecule transferring, molecular recognition and enzyme simulation selectively.

According to the requirement of film-forming for the structure, the primary hydroxyl of CD can be modified by the long alkyl chain for obtaining the modified CD consisting of hydrophilic groups (CD) and hydrophobic groups (alkyl chain). Some results demonstrate that the modified hydrophilic–hydrophobic CD is able to combine the functional groups and form multi-layers of the film of Langmuir–Blodgett (LB) accumulated by the ordered single molecular film on the surface of water. Cavities of the CD have regular arrangement, benefiting the application of membrane extraction, selective biosensor. The modified CD with –SH or –RSH at C6 is attached in the cavities of the CD through the specific adsorption capacity between sulphur and gold as well as sliver, forming the single molecular film with selectivity on the surface of metals, further leading to the changes in the properties of electrochemistry on electrodes. Therefore, this novel film can be utilized in separation technology such as chromatography.

The single molecular film derived from lipoamide-β-CD derivatives effectively inhibits the transmission of [Fe $(CN)_6$]$^{3-}$/[Fe$(CN)_6$]$^{4-}$ and selectively penetrates the complex of CD and carboxylic acid diferrocene. Simultaneously, the employment of specific objects such as borneol, ursodesoxy cholic acid, etc, competes with carboxylic acid diferrocene for combining with CD to inhibit the transmission of carboxylic acid diferrocene. By this method, the electrodes modified by CD can be subjected to detect ursodeoxycholic acid and dehydrocholic acid separated by capillary electrophoresis.

The complex consisting of azobenzene derivatives and β-CD generates the assembling film by self-assembling on the surface of metals. Compared with the self-assembling film containing 100% azobenzene, the electrochemical activity is significantly enhanced since the arrangement density is decreased by the insertion of CD into azobenzene allowing the configuration transformation and inhibiting the aggregation of azobenzene groups on the surface of metals. In addition, azobenzene is able to combine with the single molecular film of CD. After coordination, the film area of the subject molecule does not increase, illustrating the possibility of the presence of the object molecule in the cavity of the subject. The single molecular film with the encapsulated object can be transferred to the flat hydrophobic substrates by LB, such as the glass plate covered by dodecanoic acid cadmium. By this approach, the multi-layers are manufactured by repetitive operations on the substrates. The object encapsulated into multilayers has enough space to photoisomerize, for instance, encapsulated azobenzene is capable of photochemical transformation between cis–trans isomerism. Therefore, this system shows the potential on specific materials for information storage.

6.3.3. *The progress of the research of the molecular recognition and assembling of CD in china*

In the recent years, major progress has been made in the molecular recognition and assembling of CD in China. In conclusion, the following aspects are summarized:

6.3.3.1. *Research on single-modified CDs and bis (β-CD) dimers*

Focused on the bonding energy and selectivity of CD and bis (β-CD) dimers possessing functional coordinated tethers as molecular acceptor and microcosmic and macroscopic structure, the purpose of the research on CD is to identify whether the object molecule is suitable for entering into the cavity of modified CD that induces the core of the control of the molecular recognition of CD and assembling processes based on the molecular recognition of seven varieties of CD including 160 single-modified CDs and 9 varieties of bis (β-CD) dimers (totally 71); meanwhile, given that the additional binding site bis (β-CD) dimers and the length of the chain are able to control the selectivity and assembling properties of CD, the induction suitability and multi-recognition mechanism have been successfully applied in the manufacture of encapsulation of drugs, biosensor and organic nanomolecular assemblages.

6.3.3.2. *Research on CD nanolines, nanocages and polyrotaxanes*

Researchers studied the series of cylinder and spiral advanced structures of macromolecule based on the molecular recognition, resulting in the novel assembling approach for nanosupermolecule with specific topological structure with CD as a sub-unit. The results also indicated that the length of the chain can adjust the size of hydrophilic nanocages, and CD encapsulation can be utilized to directly manufacture double-nanolines, nanocages, molecular chain and polyrotaxanes etc by metal ion and condensation polymerization, constructing the unidimensional CD nanoline, double nanolines and three-dimensional CD polyrotaxanes nanocages all stable in solid phase and liquid phase. It was found that the employment of functional groups, different metal ions and some organic molecules (for instance, C60) can effectively control the topological structures and functions of molecular assemblage, offering a novel method for the nanos-tructure of the adjustable materials of synthesized acceptors and the macroscopic properties.

6.3.3.3. *Research on thermodynamics of CDs*

By measuring the thermodynamic parameters of the supramolecule system of the modified CD, it has illustrated the resource of the molecular recognition and self-assemblage of CD. Also, enthalpy contributes to the driving power for the procedure of CD combining with the object molecules and entropy is mainly responsible for the self-assemblages of CD molecule. During the combination of the object and the main body, the distinct construction transformation and extensive desolvatization effect are considered to be primary properties for the three-dimensional coordinate bond of CD.

6.3.3.4. *Research on recognition mechanism of CDs*

Research helped us to systematically understand the linkage mode and recognition mechanism of CD and calixarene-molecular acceptor with the hydrophobic cavity. The binding energy of the bonds of CD is derived from the synergistic contribution of van der Waals, hydrogen bonds and the interaction of hydrophobic bonds. However, for calixarene, it comes from the interaction of static electricity.

6.3.3.5. Research on characterization methods on CDs

Based on the structure of CD assemblages in solutions, characterization methods are developed for the investigation of self-assemblages, including electron microscopy and spectroscopy. These approaches contribute to the research of characterizations of CD assemblages effectively, along with the applications of the structures of the solutions of other supermolecule assemblages.

References

1. Szejtli, J (1998). Introduction and general overview of cyclodextrin chemistry. *Chemical Review*, 98, 1743–1753.
2. Szejtli, J (2004). Past, present, and future of cyclodextrin research. *Pure and Applied Chemistry*, 76(10), 1825–1845.
3. Emm, DV (2004). Cyclodextrins and their uses: A review. *Process Biochemistry*, 39, 1033–1046.
4. Shieh, WJ and R Hedges (1996). Properties and applications of cyclodextrins. *Journal of Macromolecular Science, part A: Pure and Applied Chemistry*, 33(5), 673–684.
5. Villalonga, R, R Cao and A Fragoso (2007). Supramolecular chemistry of cyclodextrins in enzyme technology. *Chemical Reviews*, 107(7), 3088–3116.
6. Bjerre, J, C Rousseau, L Marinescu and M Bols (2008). Artificial enzymes, "Chemzymes": Current state and perspectives. *Applied Microbiology and Biotechnology*, 81, 1–11.
7. Breslow, R (2005). *Artificial Enzymes*. Weinham: Wiley-VCH Verlag GmbH & Co. KGaA.
8. Breslow, R (1995). Biomimetic chemistry and artificial enzymes: Catalysis by design. *Accounts of Chemical Research*, 28(3), 146–153.
9. Breslow, R (1998). Studies in biomimetic chemistry. *Pure and Applied Chemistry*, 70(2), 267–270.
10. Breslow, R (2009). Biomimetic Chemistry: Biology as an inspiration. *The Journal of Biological Chemistry*, 284(3), 1337–1342.
11. D'souza, VT and MI Bender (1987). Miniature organic models of enzymes. *Accounts of Chemical Research*, 20, 146–152.
12. Breslow, R and SD Dong (1998). Biomimetic reactions catalyzed by cyclodextrins and their derivatives. *Chemical Reviews*, 8(5), 1997–2011.
13. Breslow, R, J Doherty, BG Guillot and C Lipsey (1978). β-cyclodextrinybisimidazole a model for ribonuclease. *Journal of the American Chemical Society*, 100(10), 3227–3229.
14. Breslow, R and R Xu (1993). Quantitative evidence for the mechanism of RNA cleavage by enzyme mimics: Cleavage and isomerization of UpU by Morpholine buffers. *Journal of the American Chemical Society*, 115(23), 10705–10713.
15. Breslow, R and C Schmuck (1996). Goodness of fit in complexes between substrates and ribonuclease mimics: Effects on binding, catalytic rate constants, and regiochemistry. *Journal of the American Chemical Society*, 118(28), 6601–6605.
16. Liu, L and R Breslow (2003). Dendrimetric pyridoxamine enzyme mimics. *Journal of the American Chemical sSociety*, 125(40), 12110–12111.

17. Breslow, R, JW Canary, M Varney, ST Waddell and D Yang (1990). Artificial transaminases linking pyridoxamine to binding cavities: Controlling the geometry. *Journal of the American Chemical Society*, 112(13), 5212–5219.

18. Breslow, R, PJ Duggan, D Wiedenfeld and ST Waddell (1995). The binding properties of cyclophane dimers. *Tetrahedron Letters*, 36(16), 2707–2710.

19. Dong, SD and R Breslow (1998). Bifunctional cyclodextrin metalloenzyme mimics. *Tetrahedron Letters*, 39, 9343–9346.

20. Zhao, H, FW Foss Jr, and R Breslow (2008). Artificical enzymes with thiazolium and imidazolium coenzyme mimics. *Journal of the American Chemical Society*, 130(38), 12590–12591.

21. Tamilselvi, A, M Nethaji and G Mugesh (2006). Antibiotic resistance: Mono- and dinuclear zinc complexes as metallo-β-lactamase mimics. *Chemistry — A European Journal*, 12, 7797–7806.

22. Liu, L, M Rozenman and R Breslow (2002). Hydrophobic effects on rates and substrates selectivity in polymeric transaminase mimics. *Journal of the American Chemical Society*, 124(43), 12660–12661.

23. Skouta, R, S Wei and R Breslow (2009). High rates and substrate selectivities in water by polyvinylimidazoles as transaminase enzyme mimics with hydrophobically bound pyridoxamine derivatives as coenzyme mimics. *Journal of the American Chemical Society*, 131(43), 15604–15605.

24. Hans-Jurgen, Thiem, Michael and R Breslow (1998). Molecular modeling calculations on the acylation of β-Cyclodextrin by ferrocenylacrylate esters. *Journal of the American Chemical Society*, 110(26), 8612–8616.

25. Breslow, R and N Nesnas (1999). Burst kinetics and turnover in an esterase mimic. *Tetrahedron Letters*, 40, 3335–3338.

26. Biliang, Z and R Breslow (1993). Enthalpic domination of the chelate effect in cyclodextrin dimers. *Journal of the American Chemical Society*, 115, 9353–9354.

27. Breslow, R and Z Biliang (1996). Cholesterol recognition and binding by cyclodextrin dimers. *Journal of the American Chemical Society*, 118(35), 8495–8496.

28. Breslow, R and Z Biliang (1994). Cleavage of phosphate esters by a cyclodextrin dimer catalyst that binds the substrates together with La3+ and hydrogen peroxide. *Journal of the American Chemical Society*, 116(17), 7893–7894.

29. Breslow, R, PJ Duggan and P James (1992). Cyclodextrin-B12, a potential enzyme-coenzyme mimic. *Journal of the American Chemical Society*, 114, 3982–3983.

30. Breslow, R and Z Fang (2002). Hydroxylation of steroids with an artificial P-450 catalyst bearing synthetic cyclophanes as binding groups. *Tetrahedron Letters*, 43, 5197–5200.

31. Breslow, R, X Zhang and Y Huang (1997). Selective catalytic hydroxylation of a steroid by an artificial cytochrome P-450 enzyme. *Journal of the American Chemical Society*, 119(19), 4535–4536.

32. Breslow, R, X Zhang, R Xu, M Maletic and R Merger (1996). Selective catalytic oxidation of substrates that bind to metalloporphyrin enzyme mimics carrying two or four cyclodextrin groups and related metallosalens. *Journal of the American Chemical Society*, 118(46), 11678–11679.

33. Breslow, R, Z Yang, R Ching, G Trojandt and F Odobel (1998). Sequence selective binding of peptides by artificial receptors in aqueous solution. *Journal of the American Chemical Society*, 120(14), 3536–3537.

34. Breslow, R and B Zhang (1992). Very fast ester hydrolysis by a cyclodextrin dimer with a catalytic linking group. *Journal of the American Chemical Society,* 114, 5883–5884.

35. Breslow, R, N Greespoon, T Guo and R Zarzycki (1989). Very strong binding of appropriate substrates by cyclodextrin dimers. *Journal of the American Chemical Society,* 111, 8296–8297.

36. Mugesh, G and HB Singh (2000). Synthetic organoselenium compounds as antioxidants: Glutathione peroxidase activity. *Chemical Society Reviews,* 29, 347–357.

37. Sarma, BK and G Mugesh (2008). Antioxidant activity of the anti-inflammatory compound Ebselen: A reversible cyclization pathway via selenenic and seleninic acid intermediates. *Chemistry — A European Journal,* 14, 10603–10614.

38. Mugesh, G and HB Singh (2000). Heteroatom-directed aromatic lithiation: A versatile route to the synthesis of organochalcogen (Se,Te) compounds. *Accounts of Chemical Research,* 35(4), 226–236.

39. Sarma, BK and G Mugesh (2006). Biomimetic studies on selenoenzymes: Modeling the role of proximal histidines in thioredoxin reductases. *Inorganic Chemistry,* 45, 5307–5314.

40. Breslow, R, AW Czarnik, M Lauer, R Leppkers, J Winkler and S Zimmerman (1986). Mimics of transaminase enzymes. *Journal of the American Chemical Society,* 108(8), 1969–1979.

41. Sarma, BK and G Mugesh (2008). Thiol cofactors for selenoenzymes and their synthetic mimics. *Organic and Biomolecular Chemistry,* 6, 965–974.

42. Desper, JM and R Breslow (1994). Catalysis of an intramolecular aldol condensation by imidazole-bearing cyclodextrins. *Journal of the American Chemical Society,* 116(26), 12081–12082.

43. Panda, A, G Mugesh, HB Singh and RJ Butcher (1999). Synthesis, structure, and reactivity of organochalcogen (Se,Te) compounds derived from 1-(N,N-Dimethylamino)naphthalene and N,N-Dimethylbenzylamine. *Organometallics,* 18(10), 1986–1993.

44. Mugesh, G, A Panda, HB Singh and RJ Butcher (1999). Intramolecular Se...N nonbonding interactions in low-valent organoselenium derivatives: A detailed study by 1H and 77Se NMR spectroscopy and X-Ray crystallography. *Chemistry — A European Journal,* 5(5), 1411–1421.

45. Mugesh, G, A Panda, S Kumar, SD Apte, HB Sing and RJ Butcher (2002). Intramolecularly coordinated diorganyl ditellurides: Thiol peroxidase-like antioxidants. *Organometallics,* 21(5), 884–892.

46. Rezac, M and R Breslow (1997). A mutase mimic with cobalamin linked to cyclodextrin. *Tetrahedron Letters,* 38(33), 5763–5766.

47. Phadnis, PP and G Mugesh (2005). Internally stabilized selenocysteine derivatives syntheses, 77Se NMR and biomimetic studies. *Orgnic and Biomolecular Chemistry,* 3, 2476–2481.

48. Day, BJ (2009). Catalase and glutathione peroxidase mimics. *Biochemical Pharmacology,* 77, 285–296.

49. Fujimoto, T, Y Sakata and T Kaneda (2000). The first Janus [2]rotaxane. *Chemical Communications,* 21, 2143–2144.

50. Murakami, H, A Kawabuchi, K Kotoo, M Kunitake and N Nakashima (1997). A light-driven molecular shuttle based on a rotaxane. *Journal of American Chemical Society*, 119(32), 7605–7606.
51. Pistolis, G and A Malliaris (1996). Nanotube formation between cyclodextrins and 1,6-Diphenyl-1,3,5-hexatriene. *The Journal of Physical Chemistry*, 100(38), 15562–15568.

7

USE OF CYCLODEXTRINS IN FOOD, PHARMACEUTICAL AND COSMETIC INDUSTRIES

Yao-Qi Tian, Xing Zhou and Zheng-Yu Jin

*The State Key Laboratory of Food Science and Technology,
Jiangnan University, Wuxi 214122, China*

Cyclodextrins (CDs) are cyclic oligomers of α-D-glucose units linked by α-1,4 glycosidic bonds in a donut-shaped ring [1]. CDs are non-toxic ingredients that are not absorbed in the upper gastrointestinal tract and almost metabolized by the colon microflora [2]. They, therefore, provide health benefits to people.

CDs have special properties dependant on their molecular structures. For instance, their hydrophobic cavities can encapsulate organic and inorganic molecules with smaller molecular size to form various inclusion compounds in liquid- or solid-state forms [3]; while their hydrophilic shells can generate non-inclusion complexes with larger molecular guests, such as amylose molecules and enzyme molecules [4, 5]. Mainly based on the formation of the two kinds of complexes, CDs are widely used in many areas, including foods, pharmaceuticals, cosmetics and personal care industries.

7.1. Uses of CDs in Food Products

7.1.1. *Active food packaging*

There is a growing interest in CDs used in preparing food packaging materials. The use of CDs in this field often generates several advantages: (a) CDs can reduce the residuals of organic volatile contaminants in the packaging materials, (b) the disgusting flavors released from foods can be removed during storage and (c) CDs also may control the releasing behaviors of the guests included in the packing materials for retaining high-quality and safety of food [6]. These included guests generally contain antiseptic, conserving and antimicrobial agents.

Recently, Ayala-Zavala *et al.* have assumed that there is an advantageous approach to deliver antimicrobial compounds using CDs as carriers [7]. CDs can contribute to antimicrobial delivery systems as they can generate inclusion complexes with antimicrobial and antioxidant agents and slowly release the active guests when the humidity levels increase in the headspace of the packaged fresh-cut fruits and vegetables. The released antimicrobial molecules can effectively protect the products against the microbial growth and the oxidation induced by light and oxygen. For instance, Qian *et al.* have used β-CD to complex with cinnamaldehyde to form an antimicrobial product by a thermal-sealed control method [8]. The basic conditions for the preparation obtained by one factor and response surface methodology are as follows: inclusion time, 2.5 h; inclusion temperature, $100°C$; and ratio of oil to β-CD, 1:1.75 (w/w). The releasing behavior of the antimicrobial product prepared under the optimum conditions is also estimated. The present data demonstrate that the releasing rate of the included cinnamaldehyde from the antimicrobial product is correlated with the relative humidity, but not with the temperature. Furthermore, in order to incorporate the antimicrobial product into the packing materials, β-CD is first grafted onto cellulose fibers using polyacrylic acid (PAA) as a crosslinking agent and the resultant fiber-β-CD complex is successfully used to encapsulate the antimicrobial cinnamaldehyde [9]. The resultant antimicrobial film produces a long shelf-life of bread than the pure β-CD-cinnamaldehyde inclusion complex. This suggests that the fiber-β-CD-cinnamaldehyde-based film is promising as an active food packing material for food products.

Allyl isothiocyanate (AITC, $CH_2=CHCH_2N=C=S$), a major pungent compound in plants, has a strong antimicrobial activity in its vapor form [10]. The minimum inhibitory concentrations of AITC towards nine kinds of bacteria, six kinds of yeasts and 10 kinds of molds are 34–110, 16–37 and 16–62 ng/mL, respectively. These required doses indicate that AITC is a potential natural antimicrobial agent for food preservation, including plant seeds, bread, cooked meat, fresh produce and cheese [11]. Nevertheless, there are several disadvantages to use AITC directly in the food system due to its volatility and strong odor that may affect the taste of food. The use of AITC in food packaging materials is also limited for its presence in an oil form. It is, therefore, suggested that AITC should be encapsulated by CDs before its application. As a result of the encapsulation,

AITC reveals a controlled behavior from the CDs-AITC inclusion complex for masking the strong odor, prolonging the antimicrobial time and increasing the antimicrobial effect.

In recent years, there are several reports to deal with the effect of CDs on the antimicrobial properties of AITC in a liquid-state condition [12, 13]. To clearly illustrate the effect of CDs on AITC in real food systems, a practical example has been drawn below to follow the preparation, releasing properties and use of CD-AITC inclusion complexes for the meat and baking products preservation.

A serial of AITC-based inclusion complexes with α-CD and β-CD first have been prepared using a co-precipitation method that has an advantage of easy observation of the complex formation [14]. The molar ratio of guest to CD is considered as one of the main factors for the complexes preparation among the tested factors, such as concentrations of CD, molar ratios of guest and host, reaction temperatures, the addition of ethanol and washing with water. The inclusion efficiency, the percentage of the included CD in total CD, is 85.6% for β-CD-AITC and 88.4% for α-CD-AITC inclusion complex powders under the optimum conditions. These data are similar to the theoretical maximum inclusion contents of AITC (86.3 μL of AITC for one gram β-CD and 100.7 μL of AITC for one gram α-CD).

The releasing behaviors of the included AITC from the α-CD-AITC and β-CD-AITC inclusion complexes are also investigated under different relative humidity (RH) of 98%, 75% and 50% [14]. The present results indicate that the content of the complexed AITC is rapidly reduced at the initial stage of storage and then the release becomes very slow. The releasing rate is dependent on not only the relative humidity, but also on the content of AITC included in the inclusion complexes. Furthermore, the included AITC in the α-CD-AITC inclusion compound has a much lower releasing rate than that in the β-CD-AITC sample. This demonstrates that the size of the AITC molecule is more suitable for encapsulating with α-CD than β-CD.

Avrami equation, Eq. (7.1), is generally performed to follow the rate constant of the included AITC released from the CD-based inclusion complexes under various relative humidities [14, 15].

$$R = \exp[-(kt)^n], \tag{7.1}$$

where,

R is the retention (%) of the volatile,
t is the releasing time (s),
k is the releasing rate constant and
n is a parameter representing the releasing mechanism.

Table 7.1. Kinetic parameters, n and k, for the included AITC released from the α-CD-AITC and β-CD-AITC inclusion complexes at different relative humidities [14].

RH (%)	β-CD-AITC			α-CD-AITC		
	n	$k \times 10^{-6}(s^{-1})$	R^2	n	$k \times 10^{-6}(s^{-1})$	R^2
50	0.4799	1.1383	0.9993	0.3788	0.0842	0.9972
75	0.5249	2.6900	0.9987	0.4102	0.2163	0.9988
98	0.5440	3.6166	0.9988	0.4588	0.7426	0.9997

As shown in Table 7.1, all of the n values for the CD-based inclusion complexes vary from 0.48 to 0.54 [14]. This indicates that the releasing exponent (n) of the included AITC belongs to the limited diffusion kinetics, since $n = 0.54$ represents the diffusion limited under the tested conditions. The releasing rate constant, k, reflects the releasing degree of the included AITC in the inclusion complexes. A higher k value is obtained at a higher relative humidity. These data further demonstrate that the releasing behaviors of the included AITC in the inclusion complexes are significantly affected by the relative humidity of environment.

A dynamic equilibrium behavior of the included AITC released from the CD-based inclusion complexes in a hermetical system has also been estimated by measuring the headspace concentrations of the free AITC above the β-CD-AITC inclusion complex. Compared to that of the pure AITC system, the releasing rate of the included AITC is more slow and affected by the relative humidities. The obtained data show that only an AITC headspace concentration of 61 ng/mL is determined at 50% RH and 105 h of storage time. Nevertheless, a dynamic equilibrium reaches at 90 h of storage time and 75% RH with an AITC headspace concentration of 288 ng/mL and a similar equilibrium reaches at 75 h of storage time at 98% RH with an AITC headspace concentration of 386 ng/mL. These results suggest that the interaction of the AITC and CDs is reversible and a dynamic equilibrium process occurs during the releasing of the included AITC from the CD-based inclusion complexes. Thus a modified reaction model (Eq. (7.2)) described by Li *et al.* [14] is used for exploring the releasing mechanism of the included AITC.

$$CD\text{-}AITC + nH_2O \leftrightarrow CD\text{-}nH_2O + AITC \qquad (7.2)$$

The modified model demonstrates that the behaviors of the included AITC released from the inclusion complexes in a special system are consistent with the dynamic equilibrium model for the gas–solid reaction. The releasing rate and the equilibrium concentration of the included AITC within the system, therefore, can be controlled by RH and the initial content of the included AITC, once the included AITC is used in a food packaging material.

The use of the β-CD-AITC inclusion complex in chilled beef slices has also been described previously [16]. The present data show that the ratio of the AITC to its β-CD-based inclusion complex at a 4 μL/g level can effectively reduce the numbers of *E. coli* and mesophilic aerobic bacteria in the chilled beef slices. Nevertheless, the β-CD-based inclusion complex can generate a whiter color of beef slices. It is, therefore, only appropriately used for the chilled beef slices that have no strict request with color.

7.1.2. *Inhibitory effect of CDs on the retrogradation of starch and starchy products*

7.1.2.1. *Effect of CDs on gelatinization properties of cereal starches*

The hydrophilic shell of CDs can interact with the hydroxyl groups of starch molecules, thus affecting the physico–chemical properties of cereal starch. For instance, β-CD significantly increases the amylose leaching, swelling power and the solubility of wheat starch via the possible disruption of amylose-lipid complex within starch granules [17]. The disruption has now been demonstrated in a simulated amylose-lipid (amylose-lysophosphatidylcholine and amylose-stearic acid) system [18]. Furthermore, β-CD and hydroxypropyl β-CD (HPβ-CD) increase the onset temperature (T_o), the peak temperature (T_p) and the conclusion temperature (T_c) of starch gelatinization, while reducing the enthalpy change (ΔH_g) during the gelatinization. This is interpreted by the fact that both the CDs decrease the formation of amylose-lipid complex in cereal starches by their complexation with starch lipids [18]. In a food system, the amylose-lipid inclusion complex formation or its absence is of technological interest because it directly affects the high-quality of starchy products, such as retarding bread firmness and improving structural integrity of cereal kernels during cooking [19, 20].

7.1.2.2. *Effect of CDs on retrogradation properties of cereal starches*

The inhibitory effect of CDs on the short-term retrogradation of rice starch has been reported [4, 21]. The major phenomena observed indicate that the retrogradation of rice amylose is reduced more by HPβ-CD than β-CD. HPβ-CD generates an existence of intermediate crystalline pattern (V+B) and retards the transformation from V- to B- type during the recrystallization of amylose molecules [21]. The intermediate crystalline pattern is observed, since HPβ-CD preferentially interacts with amylose to form the potential amylose-HPβ-CD complex. Furthermore, the degree of starch retrogradation is significantly lower with β-CD than glycerol monostearate (GMS). The better inhibitory effect indicates that the interaction of β-CD and starch is different from that of GMS and starch molecules [4]. In the latter starch/GMS system, the hydrophobic part of GMS is thought to be partially incorporated into starch helix to form an amylose-GMS inclusion complex [22]. Nevertheless, β-CD cannot be included in starch molecules due to its larger size, while it disrupts amylose-lipid inclusion complex [18] and may form an amylose-β-CD non-inclusion complex or β-CD-lipid inclusion complex.

β-CD significantly reduces the retrogradation of starch with a high-level amylose but does not affect the retrogradation of waxy starch with lower amylose [4]. This indicates that there is an interaction between β-CD and amylose molecules. A recent study, dealing with the effect of β-CD on recrystallization kinetics of rice amylose, reports that β-CD reduces the constant rate (k) while increasing the Avrami exponent (n) close to one (Table 7.2). These reveal that β-CD probably transforms the nucleation type of amylose recrystallization close to rod-like growth of sporadic nuclei. The occurrence of a new crystallite further confirms the formation of the amylose-β-CD non-inclusion complex.

In order to verify the formation of the amylose-β-CD non-inclusion complex, several models of amylose/β-CD interaction in periodic boundary conditions are established using HYPERCHEM 7.5 software [4]. The developed models consist of 2, 3, 4, 5, 6, 7 and 8 amylose fractions, respectively, and each fraction includes 14 polymeric glucose units in a rather stiff left-handed helix. The amylose fractions are symmetrically arranged along a circle (around the X-axis) with a radius (R) of

Table 7.2.   Avrami recrystallization kinetic parameters of amylose samples stored at 25°C [4].

Amylose samples	Avrami parameters		
	n	$k(h^{-n})$	r^2
+0.0%β-CD	0.84 ± 0.02a[a]	0.59 ± 0.01a	0.9940
+1.0%β-CD	0.86 ± 0.02a	0.54 ± 0.01b	0.9978
+3.0%β-CD	0.86 ± 0.01b	0.53 ± 0.01c	0.9980
+5.0%β-CD	0.96 ± 0.01c	0.50 ± 0.01d	0.9994

[a] Samples means with different lowercase letters in the same column are significantly different at $P < 0.05$.

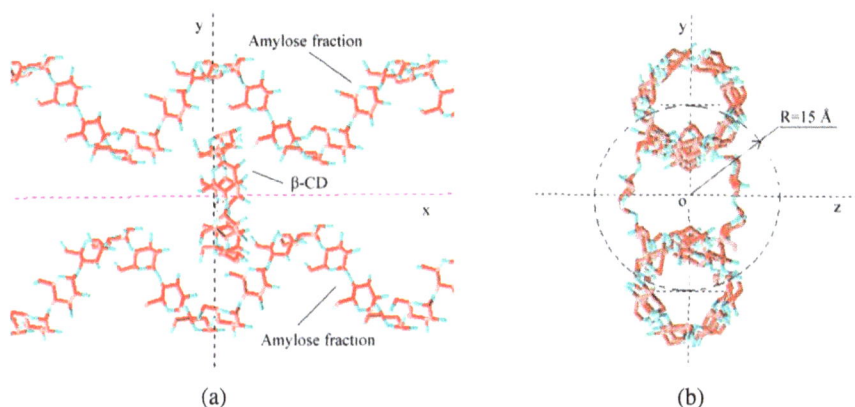

Fig. 7.1. Views of the pre-optimized conformation of two amylose fractions and one β-CD molecule along (a) z-axis and (b) x-axis direction [4].

15 Å from the origin of coordinate (the core of β-CD molecule) to the mass center of per amylose fraction. One pre-optimized model of two amylose fractions and one β-CD molecule is illustrated in Fig. 7.1. All designed models are energetically minimized using an AMBER force field by imposing a restraint on a gradient of $0.01 \times 4.186\,8$ kJ/mol·Å. The force interactions collected during the optimization show that the stability of the amylose-β-CD non-inclusion complex is primarily due to non-bonded interactions, such as van der Waals (Vdw), electrostatic force and hydrogen bond occurring in the presence of β-CD.

It has also been reported that β-CD inhibits the long-term retrogradation of rice starch [23, 24]. This inhibitory effect is mainly attributed to the interaction between β-CD and amylose-lipid complex [5]. Thus, an amylose-β-CD-lipid complex is assumed to form, based on previous reports indicating that β-CD not only disrupts the amylose-lipid complex to generate amylose-β-CD non-inclusion complex [4, 18] but also includes lipids to form an amylose-β-CD inclusion complex in starch/β-CD system. Another previous study demonstrates the amylose-β-CD-lipid complex formation using differential scanning calorimetry (DSC) and X-ray diffraction (XRD) techniques [5]. One of the possible models for the complex is optimized using an AMBER force field, as shown in Fig. 7.2. In addition, the potential amylose-β-CD-lipid complex causes disordering of starch molecules due to the short and fat aggregates observed using atomic force microscopy (AFM). These short and fat molecules are difficult to crystallize in a starch gel [25], thus resulting in the inhibitory effect of β-CD on the long-term retrogradation.

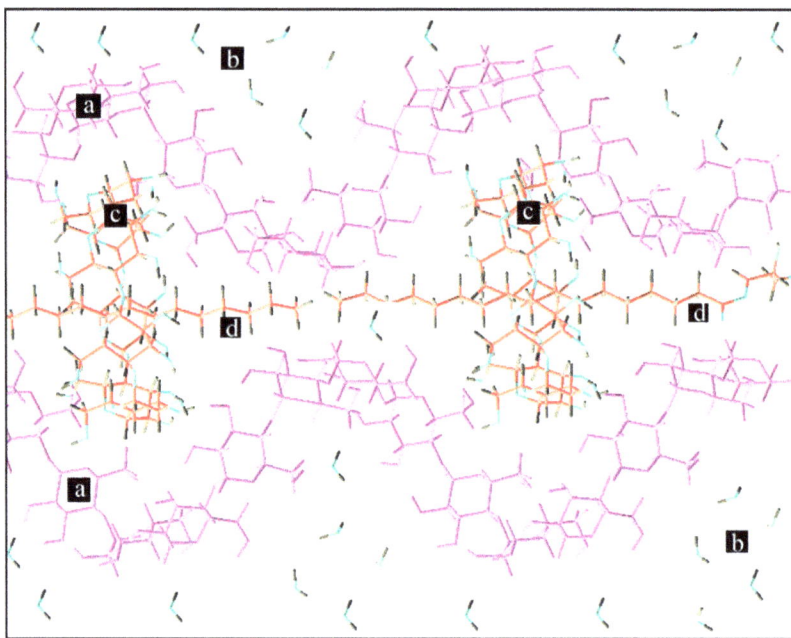

Fig. 7.2. One of the possible models proposed for the amylose-β-CD-lipid complex: (a) amylose helixes, (b) water molecules, (c) β-CD molecules and (d) monoglyceride molecules [5].

7.1.2.3. *Effect of CDs on bread staling*

Bread staling can shorten the shelf-life of bread and generate economic loss. This encourages more researchers to develop new techniques to control or retard the staling process. These techniques include, but are not limited to, the addition of glucose oxidase, xylanase, whey protein, soy protein, pentosans, arabinoxylans, hydroxypropylmethylcellulose and emulsifiers [19, 26]. Small molecule materials, such as ribose, xylose, maltose and fructose, are also considered as anti-staling agents due to their ability for the transformation of bread crystalline patterns [27].

A recent report provides a novel approach to inhibit bread staling by the addition of β-CD [19]. It has been demonstrated that β-CD reduces the hardness and lowers the change rates of the hardness, cohesiveness and springiness of bread during storage. The constant rate (k) is decreased, while the Avrami exponent (n) of bread aging is enhanced by β-CD. These data also indicate that the nucleation type of bread aging is transformed from instantaneous nucleation to rod-like growth of crystals. This transformation may be induced by a change of electrostatic property relative to the surface-activity of the surrounding β-CD molecules or starch molecules.

The crystalline patterns of bread crust and bread crumb are also affected by β-CD. β-CD causes the development of 4.4 Å and 6.8 Å peaks and a weak peak at around 5.1 Å, and develops the relative crystallinity up by 10.2% for the fresh crust. This indicates that a V-pattern is formed in the presence of β-CD. Nevertheless, β-CD retards the crystalline pattern transformation from V- to A- patterns of the aging crust via forming a V+A pattern crystallite. For fresh crumbs, an A-pattern is pronounced while it is delayed to a B-pattern by an intermediate B+V type crystallite formed due to the present β-CD [19].

7.1.3. *Protection against Oxidative Degradation, Heat-induced and Light-induced Decomposition*

CDs have been widely explored to protect most of the lipophilic food components that are sensitive to oxygen, heat and light. For instance, browning reactions of fruit and vegetable juices are inhibited by CDs, since the polyphenol oxidase is successfully removed by the complexation with CDs [28]. Other studies in recent years reported that the color of apple, pear, peach and banana were retained in the presence of natural CDs [28, 29]. Furthermore, carotenoids have been encapsulated by CDs to enhance their abilities against oxygen and light. It is reported that methyl-β-CD including carotenoids, such as zeaxanthin, lutein, lycopene and β-carotene, can be used for the purpose of cell supplementation [30]. After the inclusion complex formation, the stability of the methyl-β-CD/carotenoids complexes is better enhanced than that of the free carotenoids. Nevertheless, water-soluble inclusion complexes of the dietary carotenoid and lycopene only with α-CD and β-CD have been prepared for preventing the oxygen-, light- and heat-induced degradation.

Astaxanthin (ASX) is a well recognized carotenoid. It has wide applications as food additives, including protection against oxidation, pigmentation and reproduction taste for foods, because of its strong antioxidant and other functional properties [31]. The antioxidant activity of ASX has been demonstrated in several studies [31, 32]. In some cases, its antioxidant activity can reach to several folds stronger than that of β-carotene and vitamin E [32]. However ASX is a highly unsaturated molecule that can be easily decomposed by light and oxygen. The decomposition can cause the loss of its antioxidant properties. Another issue is its poor aqueous solubility, which limits its application as a liquid-phase antioxidant. β-CD is, therefore, used to prepare the ASX-based inclusion complex for improving the solubility and the stability of the ASX molecules [33].

Compared to β-CD, HPβ-CD is an alternative for improving water solubility of guests and has better complex forming ability [34]. Thus the HPβ-CD-ASX inclusion complex has been successfully prepared for protecting the ASX against heat-induced and light-induced decomposition [32]. The stability of the ASX molecule under oxygen and light at $4°C, 25°C$ and $50°C$ are studied, indicating that the absorbance of ASX is rapidly reduced at the initial stage and is reduced more at higher temperature than at lower temperature. In addition, the releasing rate of the included ASX is more reduced by the complexation with HPβ-CD. This decrease indicates that HPβ-CD has good capability to produce the slow releasing properties of the included ASX. The releasing mechanism by encapsulation is interpreted by an equilibrium model (Eq. (7.3)). The model reveals that there an equilibrium state occurs as the included ASX is released. The releasing behavior of the HPβ-CD-ASX inclusion complex can protect the guest of the ASX from damage caused by oxygen and light.

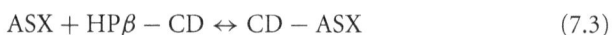

$$ASX + HP\beta - CD \leftrightarrow CD - ASX \tag{7.3}$$

7.1.4. *Encapsulation of flavors*

Flavors are of much importance in determine consumers' satisfaction and affect the consumption of foods. During production and storage processes, flavors are often reduced by oxygen and evaporation, since most natural and artificial flavors are volatile oils or liquids [35]. To inhibit the loss or the degradation of flavors during storage, it is beneficial to encapsulate volatile ingredients prior to the process of food products. These encapsulation techniques include spray-drying, freeze-drying, cocrystallization and forming inclusion complexes with CDs. Among above protections, the formation of CD-flavor inclusion complexes offers great potential use for retaining flavor materials in a multicomponent food system [29].

Many studies have been performed to use CDs for protecting flavors against heat, oxygen and evaporation [28, 29]. For example, β-CD as a stabilizing or thickening agent can retain some aroma compounds in food products [36]. As a molecular encapsulant, it also can improve the flavor quality and provide a longer period preservation compared to other encapsulants [28].

The β-CD-based encapsulation of flavors is a complexation process on a molecular scale. This process inhibits molecular interactions between the different components of natural or synthetic flavors and essential oils. The interactions thus avoided can make all flavors encapsulated into the β-CD-based inclusion complexes without changing food compositions [37]. The flavor load of these complexes varies from 6% to 15% (Table 7.3).

Table 7.3. Flavor load of CDs/flavors complexes.

Name of CD-flavor inclusion complex	Flavor load in the CD-based inclusion complex (%)	Refs.
HPβ-CD-Diallyl sulfide	13.9	38
β-CD-Garlic oil	9.4	37
β-CD-citral	9.5	37
β-CD-citronellal	9.0	37
β-CD-β-Ionone	13.3	37
β-CD-linalool	10.1	37
β-CD-bergamott oil	9.8	37
β-CD-jasmin oil	11.0	37
β-CD-sage oil	10.5	37
β-CD-cinnamon oil	10.4	37
β-CD-orange oil	8.9	37
β-CD-lemon oil	10.2	37
β-CD-lime oil	9.6	37
β-CD-onion oil	10.0	37
β-CD-mustard oil	10.2	37
β-CD-marjoram oil	9.9	37
β-CD-basil oil	10.7	29,39
β-CD-laurel leaf oil	10.8	29,39
β-CD-benzaldehyde	8.7	29,39
β-CD-caraway oil	10.5	29,39
β-CD-carrot oil	8.8	29,39
β-CD-celery oil	10.0	29,39
β-CD-Dill oil	6.9	29,39
β-CD-sage oil	8.2	29,39
β-CD-thyme oil	9.6	29,39

Natural spices with their equivalent CD-flavors complexes required to season the foods have also been reviewed in the previous studies (Table 7.4) [29, 40]. The present data indicate that the protective effect of the CD-based inclusion complexes is better than that of natural spices.

7.1.5. Elimination of disgusting tastes and unhealthy components

Reducing disgusting taste is a major factor for the consumption of various food products. Bitterness is often useful for some foods and beverages, such as coffee and beer. Nevertheless, the bitterness of citrus fruit juice is a main limitation caused by the presence of limonoids and flavonoids. Fresh citrus juice is not

Table 7.4. Equivalent amounts of β-CD-flavor inclusion complex and the corresponding natural spices determined by sensory trials.

Foods	1 g β-CD-flavor inclusion complex is equivalent to: (g)	Refs.
Onion	130–500	29, 40
Dill	150–300	29, 40
Garlic	33–100	29, 40
Cumin	3–100	29, 40
Marjoram	3–5	29, 40

bitter but becomes so depending on the pH and temperature during storage. A number of techniques, therefore, have been developed to mask or reduce the bitter components. The special one is providing CDs in the juices. The bitter taste of the grapefruit and mandarin juices is significantly reduced, while 0.3% of β-CD was added before a heat treatment of canned juices. The bitter taste of the milk casein hydrolysate, a readily digestible protein source, also can be eliminated by 10% of β-CD present in the protein hydrolysate [29].

CDs are effectively used for the removal of cholesterol from animal products, such as eggs and dairy products with the need of nutrition properties. The immobilized β-CD glass beads prepared by silanization and immobilization reaction can generate 41% of cholesterol removal in milk and a recycling efficiency of almost 100% [41]. It is also reported that the crosslinked β-CD can obviously reduce the cholesterol and retain most of physicochemical and sensory properties of mayonnaise [42]. Another previous study has been carried out to estimate the functional properties of cholesterol-removed whipping cream by β-CD, indicating that the cholesterol is almost eliminated and less time is needed for cream whipping, after the treatment of β-CD [43].

7.2. Use of CDs in Pharmaceuticals

7.2.1. *Drug solubility*

The majority of pharmaceutical agents do not have sufficient solubility in water and the conventional formulation system for insoluble drugs involves a combination of surfactants, organic solvents and extreme pH and temperature conditions. The combination generally generates irritation and other adverse reactions. CDs are not irritant and offer distinct advantages to solubilize various biomedical peptides and proteins, including interleukin-2, hormones, aspartame, tumor necrosis factor, β-amyloid peptide and albumin [44]. For instance, the solubility of cyclosporin A, an immunosuppressive agent, is significantly increased in an eye-drop form by α-CD. The increase helps the drug to penetrate into the cornea by the complexation or solid dispersion. In addition, the methylated CDs (i.e. 2,6-di-O-methyl-β-CD) with a relatively low molar substitution appear to be the most powerful solubilizer. Especially, 2,6-di-O-methyl-β-CD is a best solubilizer for cyclosporine A, since it enhances the oral bioavailability around five-fold but does not affect its lymphatic transfer [45]. Another approach of reducing the drug crystallinity on complexation with CDs also contributes to an increase of apparent drug solubility and dissolution rate. This increase is mainly attributed to the capability of CDs to form inclusion complex with drugs in dissolution mediums.

7.2.2. Drug stability

CDs are suitable to improve the stability of drugs against oxidation, hydrolysis, heat, light, metal salts and relative humidity and temperature during storage, by providing molecular shields. Table 7.5 summarizes the effects of CDs on the drug stability properties. The inclusion complex of irritating drugs in CDs also can protect the gastric mucosa for the oral route, reduce skin damage for the dermal route and mask the bitter or irritant taste and bad smell [3].

The stabilizing effect of CDs is directly related to the nature and molecular branches of the included functional groups and the kinds of CDs. Both the catalyzing effect of the nitro group and the stabilizing effect of halogen and cyanogens groups on photodegradation of 1,4 dihydropyrimidine derivatives are significantly reduced by complexation with CDs [49]. Furthermore, the

Table 7.5.   Effects of CDs on the drug stability.

Drugs	CDs	Increased effects	Refs.
Doxorubicin	HP-β-CD HP-γ-CD	Stability to acid hydrolysis and photodegradation	46
Diclofenac sodium	β-CD	Thermal stability in solid state	47
Quinaril	HP-β-CD β-CD	Stability against intramolecular cyclization	3
Digoxin	γ-CD	Stability against hydrolysis	3
Rutin	HP-β-CD γ-CD	Stability against hydrolysis	3
Paclitaxel	HP-γ-CD HP-β-CD	Stability against hydrolysis	48

photodegradation of 2-ethyl hexyl p-dimethyl aminobenzoate is reduced more by HP-β-CD in a solution than in an emulsion system. The stability constant of the inclusion complex is another main factor for the stabilizing effect of CDs on the encapsulated drugs. The lower stability constant for the low concentration of HP-β-CD/taxol system does not effectively protect the guest drugs due to a more physically unstable complex formation.

7.2.3. Drug bioavailability

A chronic treatment with peptide and protein drugs has several disadvantages, including a short biological half-life of drugs and a frequent nasal application for maintaining the therapeutic quantities of the drugs. These disadvantages require the development of drug delivery systems with the controlled-release properties. CDs act as penetration or release enhancers by improving the drug bioavailability at the biological barrier surface, i.e. skin, mucosa and eye cornea [3]. For example, CDs have been successfully applied in aqueous dermal formulations, nasal drug delivery systems, an aqueous mouthwash solution and several eye drop solutions. While for the water-soluble drugs, CDs are suitable for the increase of the drug permeability by direct interaction on mucosal membranes [50]. This is mainly interpreted by a fact that the ability of CDs to remove cholesterol can increase membrane fluidity and cause cell lysis. This process is different from a solubilization effect of detergents, since CDs can solubilize membrane components without incorporating into the membrane and the perturbing effects of CDs are mild and reversible. Other crucial factors for the improvement of drug bioavailability are the labile drug stabilization by CDs and their capability to ameliorate drug irritation [3, 50]. These advantages improve the drug contact time at the absorption site in ocular, rectal, nasal and transdermal delivery.

7.2.4. Drug safety

CDs have potential applications as drug carriers in the advanced dosage forms after most of them pass the safety evaluation. Especially, CDs have been used to reduce the drug irritation. The increased drug efficacy and potency can reduce the drug toxicity with a lower dose. β-CD increases the antiviral activity of ganciclovir on human cytomegalovirus clinical strains and the increase in the drug potency can ameliorate toxicity [51]. The toxicity of some poorly water-soluble drugs can be inhibited by a crystallization treatment. This prevents the direct contact of drugs with biological membranes and reduces their side effects [3]. Furthermore, it is well recognized that only 2%–4% of CDs are adsorbed in small intestines and the remaining section is degraded and taken up as glucose units [2]. The included drugs, therefore, can release drug components in different digestive systems. This can interpret the low toxicity observed on the oral administration of the CD-drugs inclusion complex.

7.3. Use of CDs in Daily Cosmetic and Personal Care Products

Cosmetic manufacturing needs suppressing the volatilization of perfumes, room fresheners and detergents. The major benefits of CDs are stabilizing and controlling the release of the cosmetic and personal care products from the inclusion complexes. CDs can also generate an improvement upon conversion of a liquid component to a solid form. Thus, CDs are widely applied in the main fields of skin creams, toothpaste, tissues, underarm shields, liquid and solid fabric softeners and paper towels.

The interaction of CDs with guests produces a higher Vdw force to form inclusion compounds [28]. These compounds, as stabilizers, are released slowly for prolonging the fragrance. For example, the use of CD-included fragrances in skin preparations stabilizes the fragrance against loss caused by evaporation and oxidation over a long period [52]. The antimicrobial efficiency of the products is also improved. CDs used in silica-based toothpastes increase the availability of triclosan (an antimicrobial) by complexation, thus resulting in an around three-fold enhancement of triclosan availability [53]. Furthermore, powders of dry CDs are often useful for masking odors in diapers, menstrual products and paper towels. These powders are used in hair care preparations for reducing the volatility of odoros mercaptans as well.

7.4. Conclusions

Many types of encapsulation are available to guests based on active cores and outer shells of CDs. These mainly include the formation of inclusion complexes and non-inclusion compounds. The compounds enable CDs to be widely used in food, pharmaceutical and cosmetic fields. Some of the actual or potential uses of CDs in above-mentioned areas are summarized in the relevant reviews [2, 28, 29]. Other applications of CDs still focus on a series of CD-containing products and CD-using technologies. Especially, there are growing interests to use CDs on active packaging materials for food preservation and continue the application in different areas of drug delivery and cosmetic industries. However, it is necessary to find out any possible interaction between CDs and guests, since an equilibrium reaction is often reached dependant on the native properties of the included guests and the types of CDs. Other promising applications are expected from the use of CDs in textile industry and the exploration of large-ring CDs in relevant areas, in which only very preliminary works have been published.

References

1. Lindner, K and W Saenger (1982). Crystal and molecular structure of cyclohepta-amylose dodecahydrate. *Carbohydrate Research*, 99, 103–115.
2. Szente, L and J Szejtli (2004). Cyclodextrins as food ingredients Trends in *Food Science and Technology*, 15, 137–142.
3. Challa, R, A Ahuja, J Ali and RK Khar (2005). Cyclodextrins in drug delivery: An updated review. *AAPS Pharmscitech*, 6, E329–E357.
4. Tian, YQ, Y Li, FA Manthey, XM Xu, ZY Jin and L Deng (2009). Influence of β-cyclodextrin on the short-term retrogradation of rice starch. *Food Chemistry*, 116, 54–58.
5. Tian, YQ, N Yang, Y Li, XM Xu, JL Zhan and ZY Jin (2010). Potential interaction between β-cyclodextrin and amylose-lipid complex in retrograded rice starch. *Carbohydrate Polymers*, 80, 581–584.
6. Wood, WE (2001). Improved aroma barrier properties in food packaging with cyclodextrins. In *Polymers, Laminations and Coatings Conf., TAPPI*, 367–377.
7. Ayala-Zavala, JF, L Del-Toro-Sánchez, E Álvarez-Parrilla and GA González-Aguilar (2008). High relative humidity in-package of fresh-cut fruits and vegetables: Advantage or disadvantage considering microbiological problems and antimicrobial delivering systems? *Journal of Food Science*, 73, R41–R47.
8. Qian, LL, ZY Jin and L Deng (2007). Preparation of inclusion complex of cinnamaldehyde and β-cyclodextrin by sealed thermal control method. *Food and Fermentation Industries*, 33(12), 13–16 (in Chinese).
9. Qian, LL, ZY Jin, L Deng and XH Li (2008). Inclusion and release of cinnamaldehyde by β-cyclodextrin grafted on the cellulose. *Food and Fermentation Industries*, 34(2), 16–20 (in Chinese).
10. Delaquis, PJ and PL Sholberg (1997). Antimicrobial activity of gaseous allyl isothiocyanate. *Journal Food Protection*, 60, 943–947.
11. Isshiki, K, K Tokuoka, R Mori and S Chiba (1992). Preliminary examination of allyl isothiocyanate vapor for food preservation. *Bioscience, Biotechnology and Biochemistry*, 56, 1476–1477.
12. Ohta, Y, K Takatani and S Kawakishi (2004). Effects of ionized cyclodextrin on decomposition of allyl isothiocyanate in alkaline solutions. *Bioscience, Biotechnology and Biochemistry*, 68, 433–435.
13. Ohta, Y, Y Matsui, T Osawa and S Kawakishi (2004). Retarding effects of cyclodextrins on the decomposition of organic isothiocyanates in an aqueous solution. *Bioscience, Biotechnology and Biochemistry*, 68, 671–675.
14. Li, XH, ZY Jin and J Wang (2007). Complexation of allyl isothiocyanate by α- and β-cyclodextrin and its controlled release characteristics. *Food Chemistry*, 103, 461–466.
15. Rehmann, L, H Yoshii and T Furuta (2003). Characteristics of modified β-cyclodextrin bound to cellulose powder. *Starch/Starke*, 55, 313–318.

16. Li, XH and ZY Jin (2007). Application of allyl isothiocyanate and its complex with cyclodextrin in preservation of chilled beef slices. *Transactions of the CSAE*, 7(23), 253–256 (in Chinese).

17. Kim, HO and D Hill (1984). Physical characteristics of wheat starch granule gelatinization in the presence of cycloheptaamylose. *Cereal Chemistry*, 61, 432–435.

18. Gunaratne, A and H Corke (2007). Influence of unmodified and modified cycloheptaamylose (β-cyclodextrin) on transition parameters of amylose-lipid complex and functional properties of starch. *Carbohydrate Polymer*, 68, 226–234.

19. Tian, YQ, Y Li, ZY Jin, XM Xu, JP Wang, AQ Jiao, B Yu and T Talba (2009). β-Cyclodextrin (β-CD): A new approach in bread staling. *Thermochima Acta*, 489, 22–26.

20. Biliaderis, CG and JR Tonogai (1991). Influence of lipids on the thermal and mechanical properties of concentrated starch gels. *Journal of Agriculture and Food Chemistry*, 39, 833–840.

21. Tian, YQ, Y Li, ZY Jin and XM Xu (2010). Comparison tests of hydroxylpropyl β-cyclodextrin (HPβ-CD) and β-cyclodextrin (β-CD) on retrogradation of rice amylose. *LWT-Food Science Technology*, 43, 488–491.

22. Biliaderis, CG and HD Seneviratne (1990). On the supermolecular structure and metastability of glycerol monostearate-amylose complex. *Carbohydrate Polymer*, 13(2), 185–206.

23. Tian, YQ, ZY Jin, L Deng and JW Zhao (2008). Stability and anti-retrogratation of vitamin D_3-β-cyclodextrin inclusion compound. *Journal of the Chinese Cereals and Oils Association*, 23(5), 95–98 (in Chinese).

24. Tian, YQ, XM Xu, Y Li, ZY Jin, HQ Chen and H Wang (2009). Effect of β-cyclodextrin on the long-term retrogradation of rice starch. *European Food Research and Technology*, 228, 743–748.

25. Gunning, AP, TP Giardina, CB Faulds, N Juge, SG Ring and G Williamson, *et al.* (2003). Surfactant-mediated solubilization of amylose and visualization by atomic force microscopy. *Carbohydrate Polymer*, 51, 177–182.

26. Joye, IJ, B Lagrain and JA Delcour (2009). Use of chemical redox agents and exogenous enzymes to modify the protein network during breadmaking-A review. *Journal of Cereal Science*, 50, 11–21.

27. Cairns, P, MJ Miles and VJ Morris (1991). Studies of the effect of the sugars ribose, xylose and fructose on the retrogradation of wheat starch gels by x-ray diffraction. *Carbohydrate Polymer*, 16, 355–365.

28. Martin Del Valle, EM (2004). Cyclodextrins and their uses: A review. *Process Biochemistry*, 39, 1033–1046.

29. Astray, G, C Gonzalez-Barreiro, JC Mejuto, R Rial-Otero and J Simal-Gándara (2009). A review on the use of cyclodextrins in foods. *Food Hydrocolloid*, 23, 1631–1640.

30. Pfitzner, I, PI Francz and HK Biesalski (2000). Carotenoid: methyl-bcyclodextrin formulations: An improved method for supplementation of cultured cells. *BBA-General Subjects*, 1474, 163–168.

31. Naguib, YMA (2000). Antioxidant acitivities of astaxanthin and related carotenoids. *Journal of Agriculture and Food Chemistry*, 48, 1150–1154.

32. Yuan, C, ZY Jin, XM Xu, HN Zhuang and WY Shen (2008). Preparation and stability of the inclusion complex of astaxanthin with hydroxypropyl-β-cyclodextrin. *Food Chemistry*, 109, 264–268.

33. Chen, X, R Chen, Z Guo, C Li and P Li (2007). The preparation and stability of the inclusion complex of astaxanthin with β-cyclodextrin. *Food Chemistry*, 101, 1580–1584.
34. Yuan, C, ZY Jin and XH Li (2008). Evaluation of complex forming ability of hydroxypropyl-β-cyclodextrins, *Food Chemistry*, 106, 50–55.
35. Lubbers, S, P Landy and A Voilley (1998). Retention and release of aroma compounds in food containing proteins. *Food Technology Chicago*, 52, 68–74.
36. Jouquand, C, V Ducruet and P Giampaoli (2004). Partition coefficients of aroma compounds in polysaccharide solutions by the phase ratio variation method. *Food Chemistry*, 85, 467–474.
37. Szente, L and J Szejtli (2004). Cyclodextrins as food ingredients. *Trends in Food Science and Technology*, 15, 137–142.
38. Bai, YX, B Yu, XM Xu, ZY Jin and YQ Tian (2010). Comparison of encapsulation properties of major garlic oil components by hydroxypropyl beta-cyclodextrin. *European Food Research and Technology*, 231, 519–524.
39. Szejtli, J, L Szente and E Banky-Elod (1979). Molecular encapsulation of volatile, easily oxydizable flavor substances by cyclodextrins. *Acta Chimica Science Hungarica*, 101, 27–46.
40. Lindner, K (1982). Using cyclodextrin aroma complexes in the catering. *Nahrung*, 26, 675–680.
41. Kwak, HS, SH Kim, JH Kim, HJ Choi and J Kang (2004). Immobilized bcyclodextrin as a simple and recyclable method for cholesterol removal in milk. *Archiers of Pharmacal Research*, 27, 873–877.
42. Jung, TH, HJ Ha, J Ahn and HS Kwak (2008). Development of cholesterol-reduced mayonnaise with crosslinked β-cyclodextrin and added phytosterol. *Korean Journal for Food Science of Animal Resources*, 28, 211–217
43. Shim, SY, J Ahn and HS Kwak (2003). Functional properties of cholesterol-removed whipping cream treated by β-cyclodextrin. *Journal of Dairy Science*, 86, 2767–2772.
44. Uekama, K, F Hirayama and T Irie (1998). Cyclodextrin drug carrier systems. *Chemical Reviews*, 98, 2045–2076.
45. Ichikawa, T, A Kanai and Y Yamazaki (1995). Tear secretion-stimulating effect of cyclosporin *A* eyedrops in rabbits. *Atarashii Ganka*, 12, 983–987.
46. Brewster, ME, T Loftsson, KS Estes, JL Lin and H Friðriksdóttir (1992). Effects of various cyclodextrins on solution stability and dissolution rate of doxorubicin hydrochloride. *International Journal of Pharmaceutics*, 79, 289–299.
47. Cwiertnia, B, T Hladon and M Stobiecki (1999). Stability of diclofenac sodium in the inclusion complex in the beta cyclodextrin in the solid state. *Journal of Pharmacy and Pharmacology*, 51, 1213–1218.
48. Singla, AK, A Garg and D Aggarwal (2002). Paclitaxel and its formulations. *International Journal of Pharmaceutics*, 235, 179–192.
49. Mielcarek, J (1997). Photochemical stability of the inclusion complexes formed by modified 1,4-dihydropyridine derivatives with beta cyclodextrin. *Journal of Pharmaceutical and Biomedical Analysis*, 15, 681–686.
50. Matsuda, H and H Arima (1999). Cyclodextrins in transdermal and rectal delivery. *Advanced Drug Delivery Reviews*, 36, 81–99.

51. Nicolazzi, C, V Venard, A Le Faou and C Finance (2002). In vitro antiviral activity of the gancyclovir complexed with beta cyclodextrin on human cytomegalovirus strains. *Antiviral Research*, 54, 121–127.
52. Hedges, RA (1998). Industrial applications of cyclodextrins. *Chemical Reviews*, 98, 35–44.
53. Loftsson, T, N Leeves, B Bjornsdottir, L Duffy and M Masson (1999). Effect of cyclodextrins and polymers on triclosan availability and substantivity in toothpastes in vivo. *Journal of Pharmaceutical Sciences*, 88, 1254–1258.

8

APPLICATION
OF CYCLODEXTRINS
IN NON-INDUSTRIAL AREAS

Xue-Hong Li* and Zheng-Yu Jin†

*School of Food and Bioengineering
Zhengzhou University of Light Industry
Zhengzhou 450002
†The State Key Laboratory of Food Science and Technology
School of Food Science and Technology, Jiangnan University
Wuxi 214122, China

8.1. Application of Cyclodextrins in Analytical Chemistry

The cyclodextrin (CD) molecules have secondary 2- and 3- hydroxyl groups located on the upper, broad edge and primary 6-hydroxyl groups on the narrow side of the molecule. This leads to a relatively hydrophobic cavity and permits complexation of hydrophobic portions of the guest molecules [1]. Reaction of any polar regions of the guest molecules with the surface hydroxyls combined with the hydrophobic force in the cavity provides the three-point interaction required for chiral recognition. Therefore, CDs can form inclusion complexes with a wide range of guest molecules including organic or inorganic compounds and noble gases, and have been shown to discriminate between positional isomers, functional groups, homologues and enantiomers [2–4].

These properties of CDs allow their extensive use in analytical chemistry. For example, CDs can be used for the separation of enantiomers by high performance liquid chromatography (HPLC) or gas chromatography (GC). In these cases, the chiral stationary phase or mobile phase contains CDs or their derivatives. CDs also can be used as enzyme inhibitors in affinity techniques. In spectrometry, CDs can serve as complexing agents in UV/VIS spectrophotometry, chiral shift agents in nuclear magnetic resonance (NMR) and selective agents altering spectra in circular dichroism analysis. As chiral recognition agents, CDs are of great importance in various electrophoresis techniques and thin-layer chromatography (TLC) [5, 6]. In addition, CDs have a wide range of application in microdialysis, solid- and liquid-phase extractions, separation through liquid and composite membranes and molecularly imprinted polymers.

8.1.1. *Applications of CDs in chromatographic separations*

For chiral separation, a portion of the guest molecule must enter the hydrophobic cavity and a hydrogen bonding region of the guest molecule must interact with the edge of the cavity. Therefore, the formation of inclusion complex between CDs and guests is greatly affected by the hydrophobicity, shape and size of guest molecules. Thus, CD complexation is of high selectivity, especially stereoselectivity and can give a highly selective system for chromatographic separation. The partitioning and binding of many hydrophobic and hydrophillic organic molecules to the CD cavity can be much more selective than the partitioning and binding to a single solvent or to a single traditional stationary phase. For this reason, CDs find their use in typically difficult separations of enantiomers, diastereomers, structural isomers and geometric isomers, in all current types of chromatography [7, 8]. Among the three kinds of CDs, β-CD has shown the widest application. According to the structural information, β-CD has a pronounced kink in its structure, whereas the α-CD and γ-CD are more planar. α-CD can complex single phenyl groups or napthyl groups end-on, while β-CD can accept napthyl groups and heavily substituted phenyl groups and γ-CD is effective for bulky steriod-type molecules [9].

8.1.1.1. *CDs in TLC*

Most of the separations have been carried out using HPLC with a CD-bonded phase or with CD mobile phase additives. Nevertheless, TLC is still popular as routine and spot analyses owing to its simplicity and low cost, and a large number of applications of CDs and their derivatives in TLC separations have been reported as the components of the mobile and stationary phases. In fact, a number of the positional, geometrical and optical isomers of bioactive compounds have been successfully separated by adding β-CD or its analogs in aqueous mobile phases in TLC or on β-CD-bonded silica-gel plates [10].

Aqueous CD solutions used as mobile phases in TLC have some advantages compared to traditional organic solvents to improve the selectivity or to enhance the chromatographic detection. Hinze *et al.* have studied the separation of 25 kinds of phenols and naphthols as well as 18 kinds of substituted benzoic acid derivatives by TLC with polyamide plates as solid phase and α-CD in the mobile phase. The result showed that the isomer of 0-, *m*- and *p*-nitrophenols which were easily embedded by α-CD had a larger R_f value, where the general order of R_f from large to small was para-, meta- and ortho-substituted isomes [11]. Lepri *et al.* investigated the separation of methylthiohydantoin derivatives of D- and L- amino acids and various naphthyl derivatives on SiL C18- 50F plates with β-CD as chiral agent in aqueous organic mobile phase. The result showed that the optical isomers of dansyl-, dinitrophenyl-, dinitropyridyl- and a-naphthylamide substituted amino acids had been separated completely [12].

Joseph *et al.* investigated the reversed-phase (RP) thin-layer chromatographic separations of enantiomers of dansyl-amino acids using β-CD as a chiral selector in the mobile phase. After being converted to their 5-dimethylamino-1-naphthalene sulfonyl (dansyl) derivatives, the D,L-racemates of nine proteinogenic amino acids could be separated by TLC [13].

α-, β- and γ-CDs have special application conditions due to their different cavity diameters. In general, α-CD is not fit for the selection of larger molecules because of its narrow cavity, and β-CD, which has a larger cavity diameter, proved a wider application in TLC separations. But there is a limitation in the use of native β-CD as a mobile phase additive due to its low solubility although β-CD is the least expensive. Methods to overcome the limitation of β-CD is to modify the CD synthetically or add large amounts of urea to the mobile phase so as to increase its solubility in water. Therefore, in recent decades, highly water-soluble CD derivatives or CD polymers were used widely and showed good selectivity abilities in the TLC separation of a wide variety of compounds.

Duncan and Armstrong reported the enantiomer separations of amino acid derivatives and alkaloids by TLC on different types of RP plates with a mobile phase containing maltosyl-β-CD [14]. Hydroxypropyl-β-CD (HP-β-CD) has also proved to be effective chiral mobile phase additives for the TLC enantiomeric separation of various chiral compounds, including dansyl- and β-naphthylamide amino acids. Also reported was the resolution of isomeric ortho-, meta- and para-substituted benzenes, pesticide, polycyclic aromatic hydrocarbon (PAHs) and drug test mixtures by TLC on a polyamide stationary phase with an aqueous solution of urea-solubilized β-CD as mobile phase. Huang *et al.* reported the separation of some racemic aromatic amino acids and amino alcohols by TLC on cheaper cellulose plates with mobile phases containing urea-solubilized β-CD or α-CD. They found that the compounds resolved in this experiment possessed the special molecule feature with a benzene ring and polar group in the para-position of the aromatic ring, and that the retention order of enantiomers of amino acids was opposite to that of the amino alcohol series [15]. Politzer *et al.* reported the separation of various laser dyes of the coumarin, rhodamine, bimane families and a number of dye analogs on TLC with silica gel, polyamide and C18 reverse phase plates as solid phase and aqueous β-CD, hydroxyethyl β-CD or HP-β-CD containing urea as mobile phases. The result showed that the HP-β-CD appeared to be the most effective for enhancing the dye migration [16].

The separation of two kinds of common bile acids, chenodeoxycholic acid (CDCA) and deoxycholic acid (DCA) as well as their conjugated derivatives by TLC has not been resolved for a long time. The two dihydroxylated compounds differ only from each other in the position of hydroxyl groups in the 5b-steroid nucleus. A number of TLC techniques for improving the separation efficiency have been proposed such as RP partition TLC, a high-performance thin-layer chromatographic (HPTLC) plate coated with finer particle size adsorbents, multiple developments of a TLC plate with the same or different solvent systems, two-dimensional (2D) TLC in the two different directions, a 2D TLC plate coated with both NP- and RP-adsorbents in each dimension, overpressured TLC. Although the separation efficiency of some of the recalcitrant pairs was enhanced by these approaches, the degree of resolution is still unsatisfactory [17]. Toshiaki *et al.* developed a method for the separation of the unconjugates and conjugates of CDCA and DCA by 2D RP TLC with methyl β-CD (Me-β-CD). In their study, a high degree of separation of individual bile acids in each homologous series was achieved on a RP–HPTLC plate by developing with aqueous methanol in the first dimension and the same solvent system containing Me-β-CD in the second dimension. The present method proved very simple and efficient in separating and characterizing these bile acids present in biological materials [18].

In the TLC assessment of drug purity, the drug degradation often occurs during the separation procedure. Grinberg *et al.* added γ-CD in spotting solution of TLC with a mobile phase containing hexadecyl trimethylammonium bromide, and the drug decomposition were effectively inhibited by the complex behavior of γ-CD [19].

8.1.1.2. *CDs in GC*

CDs and their derivatives could be used in GC as stationary phases for the separation of a wide variety of compounds. In these cases, the elution order of the resolved compounds in GC is in accordance with their binding ability to the CDs immobilized in the stationary resins [20]. Actually, the native CDs are scarcely used as a stationary phase due to their low separation efficiency, and the

CD derivatives are frequently used, such as (2,3,6-tri-O-methyl)-β-CD and (2,3, 6- tri-O-pentyl)-α, β-and γ-CDs.

There are some reports that polyurethane resins cross-linked with various diisocyanates were linked with the CD derivatives, and therefore utilized in GC for separating a series of compounds such as alcohols, ketones, esters, isomeric xylenes, picolines and tidines. The immobilized CDs in stationary phases included α- and β-CD acetate, permethylate, propionate, butyrate and valerate. The results proved the formation of inclusion complex between the CDs and the compounds from the gaseous phase [21]. Recently, the study on the applications of the CDs in GC has been focused on their chiral selectivity. It had been reported that the series of enantiomers such as amines and amino acid derivatives, alkylated glycerols and various lactones, cyanohydrins and carbohydrates, alkyl halides, olefins, ketones diols, triols cyclic acetals and some chiral drugs have been separated successfully using the pentylated CD derivatives as GC stationary phases [22, 23]. Celite support covered with α-CD could be used for the separation of enantiomeric mixtures of α- and β-pinene, limonene and camphene. Permethylated CDs stationary phase could resolve the racemic alkanediols, substituted carboxylic acid esters, proline methyl ester and heptamethynonane as well as various classes of volatiles [24]. Armstrong *et al.* prepared a kind of dipentyl trifluoroacetyl γ-CD (DP-TFA-γ-CD) stationary phase column for capillary GC and resolved about 120 of 150 pairs of enantiomers by the GC system, in which the γ-CD seemed to possess the widest range of application in the three kinds of CDs [25].

Usually, the stationary phase including CDs in GC was prepared by depositing the modified CDs dissolved in a dimethylformamide solution onto the support (e.g. Chromosorb W) and then evaporated the solvent by vacuum heat-treatment. Some reports described glass or fused silica capillary GC column coated with peralkylated α-, β- and γ-CDs dissolved in polysiloxanes. However, the coverage of CD in the surface is limited and results in limited preparative capacity [26]. In addition, the kinetics of inclusion are relatively slow and result in poor peak shape which also hinders the use of CDs as preparative phases.

More recently, some new kinds of modified CDs have been developed which expand the range of compounds which can be resolved and also extend their use in GC. Up to now, the GC columns based on derivatized CDs have realized commercialization. Sigma corporation had developed a series of stable derivatized CD stationary phases named DEX columns for high resolution analyses of optical and positional isomers. They prepared these functional columns by adding permethylated α-CD, β-CD or γ-CD to a phenyl-containing polysiloxane stationary cophase. The DEX columns could separate chiral compounds without derivatization, and enantiomers and positional isomers are resolved by slight

Table 8.1. Enantiomeric separations achieved with DEX columns [27].

DEX column	Content of CD	Probability of achieving separation	Compounds separated
DEX 120	20% (α-CD)	40%–50%	Alcohols, diols, epoxides, ethers, halohydrocarbons, ketones, positional isomers
DEX 110 DEX 120	10% (β-CD) 20% (β-CD)	80%–90%	Acids, amines, alcohols, diols, esters, ethers, halohydrocarbons, hydrocarbons, ketones, positional, isomers, silanes, terpenes, terpineols
DEX 120	20% (γ-CD)	40%–50%	Acids, amines, esters, halohydrocarbons, ketones, positional isomers

differences in their molecule structures involved in forming reversible inclusion complexes with the CD derivatives [27]. Some selectivity properties of DEX columns were described in Table 8.1.

8.1.1.3. *CDs in HPLC*

Since 1965, CDs and their derivatives have been applied in HPLC, although there occurs some substantial problems in the early-stage use of CDs, such as the difficulty of accessibility of the CD cavities linked to the stationary polymer, the low strength of stationary polymer gels including CDs, etc [28]. In the following years, with the development of some newer CD-phases, this area achieved great success. The new CD derivatives extends the area available for chiral interactions, they are able to separate a much wider range of compounds than native CDs, they are competitively priced and the analogs of the more popular cellulose phases are considerably more stable even allowing reverse phase operation [29]. The preparative capacity of these new phases is also better than native CDs in cases of improved chiral recognition. These CD derivatives include acetylated, S-hydroxypropylated, S or R-naphthyl ethylcarbamate, 3,5-dimethyl phenyl carbamate, cyclobond PT para-toluoylester, etc.

There are two kinds of approaches for the use of CDs in HPLC: the chiral stationary bonded CDs in which the mobile phases including highly soluble modified CDs, and the former is of far greater importance than the latter [30].

In normal phases HPLC (polar columns), the least polar component of the mobile phase have priority to be encapsulated in the CD cavity and cannot be easily replaced by another compound. The chiral separation is the result of the structure of the analyte and the competitive reactions between the stationary CDs and the components of the mobile phase. A mobile phase of water-organic liquid mixture is used in RP HPLC and the selectivity is the result of the formation of an inclusion complex of CDs with the hydrophobic part of the analyte. Sometimes, especially in some HPLC with polar organic phases, instead of the inclusion complexation, chiral separations are mainly dependent on the analyte interaction with secondary hydroxyl groups on the outer rim of the CD [31].

8.1.1.3.1. CD-bonded stationary phases

CD-bonded stationary phases are suitable for the separation of positional, geometric and optical isomers, derivatives of dansyl racemic amino acids, analogues of nicotine and nicotine analogs, arbitrates and derivatives of benzodiazepin as well as organic nitrates, imidazol derivatives, etc. Table 8.2 shows that some enantioselective stationary phases are based on modified CDs [32, 33]. The first CD derivatives used as a chiral stationary phase in HPLC were hydroxypropyl, acetylated and carbamoylated β-CD, which demonstrated better selectivity behaviors than native β-CD. The most common commercial species of CD derivatives include water-soluble methylated CD (2,6-di-O-methyl-β-CD) and HP-β-CD, then acetylated CD, carboxymethylated CD, naphthyl ethyl carbamate-β-CD,

Table 8.2. Some enantioselective stationary phases are based on modified CDs [32].

Product name	Stationary phase	Analytes
CD E	2,6-pentyl-3-butyryl-gamma-CD	Oxygenated terpenes, alcohols, epoxides
CD G	6-Methyl-2,3-Pentyl-gamma-CD	Monoterpene hydrocarbons, volatile/low temperature
CD H	2,6-Methyl-3-Pentyl-gamma-CD	Terpenes, alcohols, alkenes
CD 3P	2,6-Methyl-3-Pentyl-beta-CD	Terpenes, alcohols, alkenes
CD TM	6-TBDMS-2,3-Methyl-beta-CD	PCB, polycyclic or chlorinated aromatics, pesticides
CD TE	6-TBDMS-2,3-Ethyl-beta-CD	Pharmacopeia separations of essential oils
CD TA	6-TBDMS-2,3-Acetyl-beta-CD	Oxygenated terpenes, aromatics, low volatile
CD PM	2,3,6-methyl-beta-CD	Legacy phase for many analytes

p-toluoyl-β-CD, pyridyl ethylene diamine β-CD, phenyl carbamate β-CD and cationic β-CD. Peralkylated (methyl or propyl) β-CDs were used for the separation of racemates of pharmaceuticals, and the sulfated β-CD stationary phases were used in resolving a number of enantiomeric pairs of pharmaceuticals such as antihistamine and antidepressants, etc. The stationary phase bonded eheptakis (3-O-methyl) -β-CD and heptakis (2,3-di-O-methyl)-β-CD have been used for chiral separation of tryptophan analogues [34, 35].

Usually, the pH could influence the separation greatly. For example, at pH 7, virtually no separation of ibuprofen on β-CD is observed, whereas at pH 4.1 a good separation may be achieved [36, 37]. In general, amino acids should be chromatographed at a low pH in order to suppress the ionization of the acid groups and enhance the protonation of amine groups. Triethylamine phosphate and triethylamine acetate have proved to be very good buffers for CD columns. They are made by adjusting the pH of 0.1% triethylamine (TEA) solution either with orthophosphoric or acetic acid. Bhushan reported that HPLC resolution of thioridazine enantiomers from pharmaceutical dosage form using CD-based chiral stationary phase [38]. They chose ChiraDEX column for enantiomeric separation of thioridazine. The results showed that all of the organic modifier, concentration of buffer, pH and flow rate of mobile phase are important factors during the separation and the optimum conditions of resolution established proved a novel, rapid and reliable approach for separation and analysis of thioridazine enantiomers from commercial samples.

8.1.1.3.2. Aqueous CD solution as mobile phase

The CDs have many outstanding characteristics such as selective and reversible inclusion complexation, water solubility, light resistant, nil absorption in the full UV range and stabilities over a large pH range. These permit the use of CD in RP-HPLC systems as chiral additives to the mobile phase. The chief disadvantage of CD mobile phases is the requirement of a high CD concentration. Both α-CD and γ-CD possess relatively high solubility but β-CD has fairly low solubility. Hence more soluble methylated, carboxymethylated, cationic β-CD and other derivatives are used as mobile phase in HPLC due to the low solubility of β-CD [39, 40].

CD-containing mobile phases in HPLC have been successfully used for the separation of various isomers such as structural isomers, diastereomers and enantiomers. For example, ortho, meta, para isomers of cresol, xylene and all six isomers of nitrocinnamic acid were separated on the Lichrosorb RP-C18 column with β-CD solution as mobile phase [41]. Similar results were also obtained for ortho, meta and para isomers of nitrophenol, nitroaniline, fluoronitrobenzene, chloronitrobenzene, iodonitrobenzene, dinitrobenene, mandelic acid derivatives and enantiomers of mephenytoin, phenylalanine, α-pinene and pseudoephedrine. The CD-containing mobile phase has also been used for the separation of specific analytes from complex mixtures. These studies showed that CDs in the mobile phase could improve the separation of steroids, bile acids and their fluorescent derivative, and isomeric estrogens [42]. These highly selective chromatographic separations achieved with a CD-containing mobile phase are due to the difference in the stability constants of inclusion complexes in the mobile phase solution and to the difference in the adsorption of these complexes on the stationary phase.

In addition, the use of a CD-containing mobile phase also has some other significant advantages over the traditional organic solvent or mixed solvent systems [43]. First, the CD-containing mobile phase is safer than the organic or mixed solvent due to its non-toxicity and volatility. Second, the CD-containing mobile phase settles the solubility problems typically associated with the use of organic solvents and allows for the simultaneous separation of both nonpolar and polar solutes. Third, the use of CDs in mobile phase can enhance the chromatographic detection.

8.1.1.4. *CDs in affinity chromatography*

CDs are known to inhibit some enzymes, especially amylolytic enzymes including α-amylases, β-amylases, α-glucosidases, pullulanases and cyclodextrin glycosyl-transferases. Therefore, these enzymes can be purified by CD affinity chromatography [44, 45]. Many reports indicated that α-CD could inhibit the activity of β-amylase rather than α-amylase, based on the interaction between immobilized α-CD and the active site on β-amylase. An α-CD Sepharose 6B column was developed for separation of β-amylase from α-amylase according to the fact that the β-amylases bound to the column and the α-amylases did not. The α-amylases could pass through the affinity column while the β-amylase is retarded during the separation process. Finally, the β-amylase was eluted by the buffer including α-CD. The β-amylase purified by this step has a higher activity due to its purification [46, 47].

Similar separation behavior emerges in the case of debranching enzyme such as pullulanase, and the interaction occurring here is also at the active sites of such enzymes. For example, the starch-debranching enzymes from spinach leaf is loaded in β-CD Sepharose 6B and washed with sodium acetate buffer to remove other enzymes, then the retarded starch-debranching enzyme is released with the elution solution including β-CD. Actually, β-CD exhibit affinity for all types of amylolitic but with different degree [48]. The enzymes retarded on the affinity column could be eluted using various concentrations of β-CD.

As for α-amylase, there are some reports that some α-amylases of wheat and triticale can be purified by β-CD Sepharose 6B. But it is of interest that the interaction with the CD occurs at the raw starch affinity site rather than the active site of the enzyme, which means β-CD did not inhibit soluble starch hydrolysis but only the raw starch hydrolysis of the enzymes [49, 50].

Further investigations reveal the mechanism that CD affinity chromatography can be employed for the purification of amylolytic enzymes. It appears that CDs have no effect on the soluble starch hydrolyzing abilities, but inhibited the raw starch digestion of these amylolytic enzymes due to the enzyme binds to the CD column through an affinity site [51]. Thus, CD affinity chromatography could be used in the purification of raw starch-digesting amylases, such as raw starch-digesting α-amylases, amyloglucosidases and CGTases.

In addition, β-CD tetradecasulfate has a very strong affinity to fibroblast growth factor (FGF). Therefore, chromatographic system with a β-CD tetradecasulfate stationary with affinity was prepared for the purification of FGF. It was reported that they basic FGF could be purified by 200,000 times from rat chondrosarcoma. Recently an affinity column chromatography including immobilized A, D-bis (aminoethyl) sulfenyl-capped β-CD on the acrylonitrile-methyl acrylate copolymer was developed, which had properties of both carbonyl recognition and hydrophobic recognition, and could be used for the separation of compounds of similar structures having a hydrophobic site as well as a carbonyl group [52].

8.1.2. CDs in electrophoresis

It was 1982, when CDs were first used in isotachophoretic analysis of alkali and alkaline metals, in which as a complexing additive α-CD obviously improved the separation effect. Since that time, many investigations of CDs application in various types of electrophoresis have been carried out.

In capillary electrophoresis (CE), CDs and their ionic and neutral derivatives have been successfully used as additives in the carrier system for the separation of structural isomers and structurally related compounds [53]. The commonly used neutral CDs are the native α-, β- and γ-CDs and the dimethyl, trimethyl, hydroxyethyl and hydroxypropyl forms [54]. The charged CDs are carboxymethyl, sulfobutyl ether, sulfated and amino CDs. The methyl derivatives of the CD are effective in separating chiral compounds, enantiomers of terbutaline, ephedrine and carnitine. The neutral derivatives of hydroxyalkylated β-CD and the mixture of neutral and ionic CD derivatives are also frequently used for enantiomeric separations as a chiral selector. Randomly substituted sulfated CD have been successfully used for both the enantioseparation of 56 kinds of drugs including anesthetics, antiarrhythmics, antidepressants, antihistamines and antimalarials, and the chiral separation of catecholamines such as norepinephrine, epinephrine, dopa and their precursors, phenylalanine and tyrosine [55, 56]. Recently, some newly synthesized derivatives of CDs have been developed such as tertbutyl-α-, β- and γ-CDs, and could be used for the separation of the halogen derivatives of benzoic acid by CE [57].

CDs could be used as leading electrolyte additives in isotachophoresis for improving the selectivity [58]. The addition of α-CD in the leading electrolyte contributed to the complete separation of compounds such as nitrite and nitrate ions, cyanate, thiocyanate and selenocyanate ions, chlorate and perchlorate ions. CDs were also successfully used as leading electrolyte additives in the capillary isotachophoretic separation of positional isomers, such as 2-, 3- and 4-amino phenols, 1,2-, 1,3- and 1,4-diaminobenenes, and substituted aromatic sulfonic acids.

Micellar electrokinetic chromatography (MEKC), a modified CE, enabled the separation of electrically neutral analytes. However, highly hydrophobic compounds including PAHs, corticosteroids, fat-soluble vitamins and polychlorinated biphenyl (PCB) congeners could not be separated by MEKC in electrophoretic medium of sodium dodecylsulfate (SDS) [59]. The use of CDs to the SDS solution can remarkably improve the resolution of analytes. For example, a mixture of water-soluble and fat-soluble vitamin or esterom can be successfully separated by MEKC by the addition of γ-CD in the SDS electrophoretic medium [60].

8.1.3. *Applications of CDs in spectroscopy*

The high electron density existing inside the CD cavity can grip the electrons of the encapsulated guest molecules, and result in a various spectral properties of both the guest and host [61]. This effect of CDs on the spectral properties of guest molecules has led to their utilization in various spectrometric analyses, such as UV/VIS spectrophotometric analysis, fluorescence and luminescence spectroscopy and NMR spectroscopy.

8.1.3.1. *CDs in UV/VIS spectrophotometry analysis*

CDs could effectively change UV/VIS spectral properties of guest by complexing the guest. Usually, the intensity and position of the absorption bands in the spectrum are varied after the formation of an inclusion complex. Due to the hydrophobic cavity of CDs, the spectra of the included guests in aqueous solutions are very similar to that of guests in organic solvents. In UV/VIS spectrophotometry, CDs are mainly used to improve the solubility and stability of colored compounds or to increase the sensitivity and selectivity of color reactions. Therefore, CDs as a reagent could improve the sensitivity of determinations in UV/VIS spectrometry and be useful auxiliaries in the spectrophotometric determinations of a wide variety of compounds and elements [62]. CDs have been further used to calculate the dissociation constants with the Scott equation or the Benesi–Hildebrand equation owing to the remarkable spectral changes in the UV/VIS spectra upon adding CDs.

It has been reported that β-CD could improve the selectivity of the color reactions of various metal ions with triphenylmethane, xanthene acid dyes and some other coloring reagents. The effect of β-CD on the association compound system of metal (Mo, Zn, Co)-thiocyanate basic dyes such as malachite green, crystal violet, rhodamine B, rhodamine 6G and butylrhodamine B, has been investigated and the result shows that β-CD could contribute to a more sensitive and stable system which improve the solubility of the basic dyes and produce a favorable microenvironment for the color reactions [63]. β-CD could be employed to solubilize the 1, 2-amino anthraquinone in water due to the formation of inclusion complex which acts as a ligand for metal ions could be used for the determination of palladium at trace levels by spectrophotometry. In the spectrophotometric determination of microamounts of Zn based on the Zn-dithizone color reaction, β-CD could increase the apparent molar absorptivity at 538 nm by 8.37 times. In the presence of α-CD, the determination sensitivity of copper in leaves based on the color reaction of Cu(II) and mesotetrakis (4-methoxy-3-sulfopheny1) porphyrin was enhanced by 50% in the spectrophotometric analysis [64, 65].

CDs also can be used as stabilizers for color indicators to increase the stability of indicators used for the spectrophotometric determination of hydrogen peroxide in body fluids. In addition, CDs and their derivatives also have applications in enzyme assays and enzyme activity measurement. For example, glucosyl- or maltosyl-α-CD have been used to increase the accuracy and sensitivity of the assay of amylase.

8.1.3.2. *CDs in luminescence spectroscopy*

Molecular luminescence spectrometry, especially molecular fluorescence spectrometry, has become established as a routine technique in many analytical applications. In many cases, molecular luminescence spectrometry can yield a lower detection limit and greater selectivity than molecular absorption spectrometry. However, although most compounds exhibit strong fluorescence or phosphorescence in organic solvents, the intensity of luminescence is rather weak in water. Adding CDs, which form inclusion complexes with analyte molecules in aqueous solution, can result in significant enhancement of the fluorescence or phosphorescence [66]. The following are the advantages of the formation of analyte molecules inclusion complexes with the CDs [67]:

i. The structural conformation of CDs could protect the fluorescing singlet state or the phosphorescing triplet state of the analytes from external quencher.
ii. The CDs complexation could hinder the rotation of the guest molecule and decrease the relaxation of the solvent molecules, which both produce a decrease in the vibrational deactivation.
iii. The CDs cavity could provide an apolar microenvironment in which the favorable polarity and acid/base equilibria are established for enhanced quantum efficiencies and hence the intensities of luminescence.
iv. The CDs can improve the detection limit for analytes by increasing their solubility.

Inclusion complex formation with CDs usually results in a higher fluorescence quantum yield or the lifetimes of the excited states. It has been found that the fluorescence intensities of many compounds, including pyrene, various illicit drugs, narcotics, hallucinogenic, and polychlorinated biphenols are significantly increased by the complex formation with CDs and their derivatives [68]. 1-anilinonaphthalene-8-sulfonaties, which show very weak fluorescence in water, can exhibit strong fluorescence upon adding CDs into the aqueous solution and the fluorescence intensity of this compound in β-CD solution is increased by 10 times. As for the compounds such as ammonium 7-fluorobenzo-2-oxa-l, 3-diazole-4-sulfonatleabeled glutathione, acetylcysteine and some dansylated amino acids, aqueous CDs increase the fluorescence emission about eight-fold in comparison with the original values [69].

The fluorescence intensity of naphthalene in aqueous solution decreases upon aeration. However, the quenching of naphthalene by aeration is totally suppressed in the presence of a water-soluble sulfopropylated β-CD [70]. Similarly, the quenching of halonaphthalene phosphorescence in water by $NaNO_2$ can be substantially inhibited by β-CD. The rate of inhibition depends on the bond-tightness between the analyte and the CD. Retinal, which is normally insoluble in water and is not fluorescent in solution at room temperature, emits luminescence in the region of 450 nm and permits fluorescence detection when incorporated by β- or γ-CD even in air-saturated aqueous solution [71]. In the luminescence detection of volatile compounds, CDs can be used as solid matrices which efficiently trap the volatile compounds for obtaining room temperature fluorescence (RTF) and room temperature phosphorescence (RTP) from the absorbed compounds [72]. This approach was used for the determination of polynuclear aromatic hydrocarbons, nitrogen heterocycles and bridged biphenyls with subpicogram detection limits and well-resolved spectra.

The CDs also can be used to enhance chemiluminescence of the luminol related compounds. It was found that CDs were capable of increasing the light output by factors up to 300 in aqueous peroxyoxalate solution. The enhancement could be attributed to increases in reaction rate, excitation efficiency and fluorescence efficiency of the emitting species.

In most cases, the present of CDs will enhance the luminescence. However, CDs can also selectively quench the luminescence of some compounds. A study of the effect of β-CD on the fluorescence of xanthene dyes, coumarins and pyromethene-difluoroboron complexes in aqueous solution shows that β-CD enhances the fluorescence of 7-hydroxycoumarin and coumarins, but quenches the fluorescence of the 7-hydroxy-4-methylcoumarins [73]. This behavior of CDs provides a new approach to multicomponent fluorometric analysis and indicates that CDs can be used for differentiating the structures of similar compounds such as the positional isomers by the selective incorporation of the analyte.

Recently, the supramolecular systems of sensitizers mesotetrakis (4-sulfonatophenyl) porphine (TPPS4), its zinc(II) and palladium(II) complexes and mesotetrakis (4-carboxyphenyl) porphine with native CD and HPβ-CD have been studied. The results indicate that there are strong noncovalent host–guest interactions between selected sensitizers and CDs (association constants K 107 M^{-1}), and the complexation significantly effects the physical and photophysical properties of these compounds [74]. The presence of CDs prolongs the lifetimes of the triplet states of these sensitizers in monomeric form, the rate constants of quenching of the triplet states by oxygen decrease and the intensity of the fluorescence emission of the sensitizers is changed. The quantum yields of the triplet states of these sensitizers and $^1O2$ formation remain almost unchanged upon binding to CDs. In the ground states, CDs affect the acid–base equilibrium and aggregation of free porphyrin. The ability of CDs to maintain the monomeric, photodynamically active form of a sensitizer is highly important with respect to the tendency of porphyrin to aggregate at higher concentrations and ionic strengths in aqueous solutions, because the aggregates have very low or no quantum yields of singlet oxygen.

8.1.3.3. *CDs in NMR spectroscopy*

In NMR spectroscopic analysis, CDs are mainly used as chiral NMR shift reagents. The complexation with CDs can affect the character of the NMR spectra of the guests. In many cases, the effect of CDs inclusion complex formation on the NMR features of the two enantiomers of a chiral compound differs in chemical shifts [75, 76]. The ^{1}H NMR, ^{13}C NMR, ^{15}N NMR spectroscopy could be used for studying the CDs inclusion complexes and their properties, such as the structures of the inclusion compounds formed, the interaction of CDs with acids, aliphatic amines and cyclic alcohols, and optical purity of guests. One-dimensional and 2D ^{1}H NMR spectroscopy have been used to investigate the chiral recognition process in CE. The changes in the 3H and 5H protons in ^{1}H NMR spectroscopy prove the formation of an inclusion complex. When the guest molecule is enclosed in the CD cavity, the resonance signals of protons located inside the cavity (3H and 5H) are shifted in the spectrum. The signals of protons located on the outer side of the cavity (2H, 4H, 6H) remain relatively unaffected.

Up to now, NMR spectroscopy has become the most powerful approach for the study of inclusion complex formation between CDs and a variety of guest molecules, from initial 'H NMR in solution to ^{13}C NMR, ^{15}N NMR, ^{19}F NMR and ^{31}P NMR in the solid state.

A ^{19}F NMR study on the formation of diastereoisomeric inclusion complexes between fluorinated amino acid derivatives and α-CD in 10% D_2O solution shows that the chemical shifts of the D-amino acid derivatives included by α-CD are upfield from those of their L analogues [77]. The shift difference between the diastereoisomers formed with D and L enantiomers can be used for chiral analysis and optical purity determinations. For example, the interaction of β-CD with propanolol hydrochloride produces diastereomeric pairs. The protons of the antipode give ^{1}H NMR signals which differ in chemical shifts in D_2O solution at 400 MHz. The intensity of the resonance signals for each diastereoisomer has been used for optical purity determination. By adding racemate to pure ($-$) isomer, this technique is able to measure optical purity of propanolol hydrochloride in water down to the level of 1%.

8.1.4. *CDs in electrochemistry*

From 1953, much research has been done on the electrochemical behavior of CDs and their inclusion complexes. It is known that the effect of CDs on electrochemical properties of the guest molecules can be used in potentiometry, polarography and voltametry, cyclic voltammetry and amperometry. The ability of CDs to bond, orient and separate molecules and to form inclusion complexes in solution or on modified electrodes can be utilized for electrocatalysis, electrosynthesis and electroanalysis [78].

8.1.4.1. *Electrochemical behavior of CDs and CD inclusion complexes*

The addition of CDs can affect the reduction processes of organic and inorganic molecules. Many research works have been devoted to the effect of inclusion behavior of CDs on electron transfer in reversible redox systems and especially on complexation/decomplexation during oxidation or reduction processes. For example, CDs can affect the polarographic reduction wave of atrazin, and the final effects depend on the size of the CD cavity and increase with the increasing CD concentration [79]. CDs exhibit adsorption–desorption peaks on cyclic voltammograms, demonstrating adsorption processes although they don't form direct current polarographic waves. CDs decrease the surface tension and the drop time of mercury by complexation. CDs complexation depends on the potential applied electrode and exhibits a very complicated character due to 2D condensation of CDs and reorientation effects in the adsorbed state. At less negative potentials, CD molecules are oriented with the cavity perpendicular to the electrode surface, while at more negative potentials, orientation is intermediate between "parallel" and "perpendicular". These adsorption effects have been applied in the quantitative assays of CDs.

8.1.4.2. *Use of CDs in electrochemical analysis*

Recently, a regioselective electrode system with a poly (perfluorosulfonic acid)-coated electrode based on CDs complexation for the determination of *o*-nitrophenol in the presence of *p*-nitrophenol has been developed [37]. The *p*-nitrophenol shows an extraordinarily small reduction peak on a regioselective electrode in α-CD solution, whereas there is no effect of α-CD on *o*-nitrophenol. This system is 33 times more sensitive to *o*-nitrophenol than to *p*-nitrophenol and provide an accurate determination of *o*-nitrophenol in the presence of its para isomer.

Voltammetric sensors based on host-guest molecular recognition, which is responsive to anionic guests, have also been developed. These voltammetric sensors were constructed with membrane assemblies of lipophilic CD polyamine containing anion receptors deposited directly on glassy carbon electrodes with the Langmuir–Blodgett method, in which CD polyamine can bind with anionic guests in multiprotonated forms [37, 78]. The response to the anionic guests appears with the decrease of peak height in cyclic voltammetry using $[Fe(CN)_6]^{4+}$ as marker ion. The order of selectivities for positional isomers of phthalate was *m*-isophthalate, *p*-terephthalate and *o*-phthalate due to the host–guest interaction involving in the CD cavity.

A potentiometric enantioselective membrane electrode with bonded 2-hydroxy-3-trimethyl-ammoniopropyl-β-CD has been used for the enantio-purity determination of several L-enantiomers. The β-CD-ferrocene inclusion complex has been used for the amperometric determination of ascorbic acid. Alkylated α-, β- and γ-CDs have been incorporated in an ion-selective electrode as neutral ionophores and this electrode was used for the potentiometric andamperometric determination of the tricyclic antidepressives and their hydrochloride salts [80]. Cyclic voltammetry was used for measuring the association constants of inclusion complexes.

8.2. Application of CDs in Environmental Problems

CDs have a low polarity cavity in which organic compounds of appropriate shape and size can form inclusion complexes. This unique property provides CDs with a capacity to significantly increase the apparent solubility of low polarity organic compounds [81]. As a consequence, they can play a major role in environmental science in terms of solubilization of organic contaminants, enrichment and removal of organic pollutants and heavy metals from soil, water and atmosphere.

CDs have the abilities of promoting or decelerating degradations of hazardous pollutants discharged to the aqueous environments. CDs have been proposed as an alternative agent to remove the organic pollutants from contaminated sites and have the potential use for *in situ* flushing in environmental remediation. In addition, CDs do not pollute the environment due to their biodegradability which depends on the degree of substitution, and the more the substituents on the CD ring, the slower is the biodegradation. There is a research report, involving the biodegradation of eight kinds of CDs in four different soils which showed that the CDs were all biodegraded by soil microorganisms in the order of α-CD $\approx$ β-CD $\approx$ γ-CD > Ac-β-CD > HP-β-CD > peracetyl α-CD $\approx$ peracetyl β-CD $\gg$ random methylated β-CD (RM-β-CD) (Table 8.3) [82]. According to the above study, RM-β-CD is a good additive for enhancing the bioremediation because of its highest dissolution effect as well as its optimal degradation rate in the contaminated soil. RM-β-CD has a relatively slow degradation rate enough to ensure its necessary concentration for eliminating the pollutants from the soil.

8.2.1. *Removal of contaminants from waste gases*

8.2.1.1. *Sorption of organic solvent vapors from waste gases of chemical industry*

Volatile organic compounds (VOCs) are among the most common air pollutants emitted from chemical, petrochemical and allied industries, and includes most

Table 8.3. Percent biodegradation and half-lives of CDs in soil after 178 d and 280 d [82].

Test item	Biodegradation (%)		Half-life time (d)
	After 178 d	After 280 d	
Cellulose	102 ± 11	108 ± 12	35
α-CD	89 ± 3	n.m.	17.5
β-CD	94 ± 6	n.m.	17.5
γ-CD	89 ± 2	n.m.	20
Peracetyl α-CD	97 ± 4	n.m.	62
Peracetyl β-CD	96 ± 1	n.m.	65
Partially acetylated β-CD	85 ± 1	103 ± 5	17.5
HP-β-CD	72 ± 1	98 ± 6	122
RM-β-CD	-19 ± 3	n.m.	—

solvent thinners, degreasers, cleaners, lubricants and liquid fuels. Emission control of VOCs has become a major concern in air pollution prevention [83]. New regulations regarding VOC emissions demand more efficient and less costly technologies. Adsorption is a procedure chose for treating industrial waste gases, and is a useful tool for protecting the environment, in which the key is the choice of a suitable liquid absorbent. Absorption is used to remove VOCs from gas streams by contacting the contaminated air with a liquid solvent and the soluble VOCs will transfer to the liquid phase. In the case of hydrophobic VOCs, absorbents which can improve the solubility of VOCs will be required to replace water solution. CDs are regarded as useful sorbents for the removal of aromatic hydrocarbons from the gas phase.

In a research paper, six kinds of CDs including α-CD, β-CD, HP-β-CD, RM-β-CD, low methylated-β-CD (Me-β-CD) and a sulfobutyl-ether-β-CD (SBE-β-CD) were compared and evaluated in the effectiveness and the regeneration with toluene as the target hydrophobic VOC [84]. The results showed that all CD derivatives tested were able to decrease the volatility up to 95% depending on both of CD species and concentration. The absorption capability of β-CD is 250 times larger than water. The absorption efficiency is not totally correlated with static experiments, suggesting that, besides absorption constants and stability of inclusion compounds, toluene diffusion into such CD solutions is an important factor.

Because of the application limitation of water solubility of CDs, there has been considerable interest in the preparation and application of crosslinked CD polymers. The most frequently used crosslinking agent is epichlorohydrin (EP). A study concerning CD-EP polymers which were used to trap VOCs from gases was carried out, in which several polymers were prepared using low methylate β-CD and EP with various ratios, and the adsorption capacities of the obtained polymers towards toluene as a model aromatic compound had been investigated [85]. The result showed that the polymers with lower EP/Me-β-CD ratio have a higher absorption capability, whose values close to the Me-β-CD, maybe due to the better accessibility of the CD cavity.

There was also a research involving an approach for extraction of PAHs from air using solid CDs. A comparison experiment in which β-CD is replaced by α-CD provides evidence that β-CD removes PAHs vapor by formation of inclusion complexes rather than by association or adsorption interactions. Thereby, β-CD can incorporate PAHs vapor and reduce their volatilities, and can be proposed for improving ambient air quality. In this application of CDs, fluorescence and absorbance spectroscopies could be used to examine the variables that affect the formation of the PAH complexes with the solid CDs.

8.2.1.2. *Sorption of radioactive iodine from the waste gases of nuclear power plants*

It is necessary to install air filters in the running nuclear power plants with the special aim to adsorb effectively radioactive iodine vapors emitted upon malfunctions of the nuclear power plants. The normal or minor malfunction of nuclear power plants may result in the generation of considerable amounts of radioactive elemental and small amounts of organic iodine vapor or dust. The iodine waste, due to being fission intermediates, is the primary target for effective entrapment and immobilization to prevent the spreading of nuclear contaminants. If iodine can be effectively immobilized, then the majority of radioactivity remains localized.

CDs can form inclusion complexes with iodine, which makes them candidates for iodine-sorption from nuclear waste gases. In model experiments, it was shown that the aqueous solutions containing CDs and crosslinked CD polymers were selective and effective iodine absorbers [86]. Especially the α-CD derivatives (methylated and crosslinked) have high sorption capacity. On the basis of the results, the binding of elemental and organic iodine emitted into the air by chemical and nuclear power plants can be made effectively by immobilizing iodine vapor in aqueous CD solutions or in CD polymer gel beds. Such new sorbents can be employed in the air filtration systems (Table 8.4).

Table 8.4. Iodine binding performance of CD polymer filled columns as air filters [86].

Packing	Breakthrough time (h)	Iodine sorbed by the packing (mg/cm^3)	Iodine sorbed by the packing (mg/g dry sorbent)
α-CD Polymer (dry)	1	1.2	2
α-CD Polymer wetted by water	9.5	23.8	119.1
α-CD Polymer wetted by 0.1 N KI solution	18	44.4	226.8
β-CD Polymer (dry)	1	0.8	1.3
β-CD Polymer wetted by water	3	7.4	36.8
β-CD Polymer wetted by 0.1 N KI solution	16	38.6	192.8
Dextran polymer dry	<0.01	<0.2	<1
Dextran polymer wetted by water	0.2	0.8	4.3
Dextran polymer wetted by 0.1 N KI solution	2	5	25

8.2.2. *Removal of contaminants from waste water and groundwater*

The pollution of groundwater has become an environmental and economic hazard as a result of waste spillage and industrial applications such as pesticides in agriculture. Conventional treatment techniques such as sand filtration, sedimentation, flocculation, coagulation, chlorination and activated carbon are not very effective in reducing the concentration of the organic contaminants. So, there are demands for new efficient technologies for water purification. It is well known that the structure of CDs give rise to a remarkable capacity to form inclusion complexes in solution with organic molecules by host-guest interactions. Therefore, CD complexation is a procedure of choice for depollution techniques. The following examples concern the application of CDs in decontamination from waste water.

Malathion, even in a low dose, would produce a very harmful effect on environmental safety. The possible applications of α-CD in remediation of malathion contamination was assessed and the results showed that α-CD had a remarkable effect on the hydrolysis of malathion in natural groundwater. The facilitative effects increase with increasing the concentration of α-CD [87]. Several crosslinked CD gels were prepared and proposed for removal of various aromatic derivatives, such as chloro and nitro phenols, benzoic acid derivatives, dyes *et al.*, from aqueous solutions. Results of adsorption experiments showed that these crosslinked polymers exhibited high sorption capacities, and usually the control of the crosslinking reaction is the key parameter to improve the sorption properties of the material. For example, new hydrophilic CD adsorbents, which were prepared by radical copolymerization in water using 2-hydroxyethyl methacrylate or vinylpyrolidone as co-monomers, showed high sorption capacities toward benzene derivatives. The CD-carboxymethylcellulose gels synthesized with epichlorohydrin (EPI) exhibited effective extraction of β-naphtol in which the presence of carboxyl groups in the polymer networks lead to significant improvement of the sorption properties. A novel CD/EPI copolymer developed recently has the ability to extract trace aromatic compounds in water and the high extraction efficiency with recoveries between 90% and 100% for aromatic compounds at 0.02–1.67 ppm. By stirring 5 mg/ml of highly crosslinked CD/EPI copolymer in a 0.2 mmol Bisphenol A (BPA) solution for 2 h, more than 98% of the BPA was removed [88].

Batch adsorption experiments have been carried out for the removal of three basic dyes, namely C.I. Basic Blue 3, Basic Violet 3 and Basic Violet 10, from aqueous solutions using a CD polymer (CD/EPI) [89]. The results showed that the polymer adsorbent exhibited high sorption capacities toward basic dyes. Further analysis of experimental data indicated that the sorption kinetics followed a pseudo-second order equation, suggesting that the rate-limiting step may be chemisorption. Both Freundlich isotherm and Langmuir isotherm gave a good correlation for the equilibrium isotherm adsorption of basic dyes. On the basis of the Langmuir analysis, the maximum adsorption capacities were determined to be 53.2, 42.4 and 35.8 mg/g for C.I. Basic Violet 3, C.I. Basic Blue 3 and Basic Violet 3, respectively. The differences in adsorption capacities may be due to the effect of dye structure. The negative value of free energy change indicated the spontaneous nature of adsorption.

Recently, a new series of crosslinked CD polymeric nanospheres with different sizes have been synthesized successfully by a unique method of miniemulsion polymerization. These CD nanospheres exhibit a high ability to absorb aromatic organic molecules, especially aromatic rings such as toluene and phenol. Among these results, the interest is that the hazardous organic contaminants can be reduced to very low levels in waste water by the nanosphere particles [90]. The advantage that polymeric nanospheres have over inorganic materials is that they offer flexibility in processing, and can be fabricated into granular solids, fine powders, thin films and possibly smart membranes. Such flexibility enables these materials to be used for multiple applications and formats, thereby accommodating different water treatment configurations and needs. But there are also some disadvantages of CD nanospheres compared to activated carbon. For example, the CD nanospheres do not absorb moisture from air, else they lose their effectiveness.

In the environmental field, the pollution from various kinds of synthetic dyes pose serious problems besides of that from PAHs and PCBs. Synthetic dyes are manufactured and used for numerous industrial applications, such as textiles, leather goods, food manufacturing and other chemical uses. It is estimated that out of the total amount of dyes in use, 1%–2% in manufacturing and 1%–10% are released into water, air and soil. Since the interactions between the pollutant and CDs are driven mainly by hydrophobic interactions, the CD polymers, used as sorbents for the removal of various dyes, could be easily regenerated after saturation, and the organic solvents are good candidates for the regeneration of the material. The polymers were easily regenerated using ethanol as the washing solvent and the sorption capacity value remained unchanged after this wash treatment.

In addition, dye pollutants could be bleached under visible irradiation through photosensitized degradation on a semiconductor surface. But the degradation is usually inefficient due to only an adsorbed dye molecule injecting the charge from its excited state to the semiconductor's conduction band. β-CD has a stimulative effect on the TiO_2 photochemical properties because β-CD plays electron-donating and hole-capturing roles when linked to TiO_2 colloids, which lead to charge-hole recombination restriction and photocatalytic efficiency enhancement. Therefore, a kind of β-CD grafted TiO_2 hybrid powder was synthesized by a photoinduced self assembly process which exhibited high reactivity for dye pollutant degradation under visible irradiation and simulated solar irradiation [91]. The initial disappearance rate of Orange II used in this experiment increased by 2–7 times. The $\cdot O_2^-$ mediated oxidation reactions were dominant for Orange II degradation under visible irradiation and the reaction between β-CD and diffusion mediated $\cdot O_2^-$ was negligible, which makes TiO_2/β-CD an effective and stable catalyst. β-CD could increase the excited states lifetimes of unreactive guests and facilitate electron transfer from the excited dye to TiO_2 conduction band, which enhances dye pollutant degradation.

8.2.3. *Remediation of contaminated soils*

Low bioavailability of hydrocarbon pollutants can limit the biodegradation by indigenous micro-communities in soils. CDs enhance desorption of the non-polar contaminants from the solid surface and transfer them to the water-phase biofilms, where the hydrocarbon-degrading microbes work. Therefore, this special character of CDs and their derivatives can be used for enhanced removal of hydrocarbon contaminants from soil [92]. This is so-called CD-enhanced pump and treat technology for the removal of dense, non-aqueous phase liquid from the saturated soil. For example, HP-β-CD solution was added into the source zone and pumped out successfully with contaminants on the subsurface, which reduced remarkably the remediation time compared with the conventional pump and treat technology.

PAHs are pollutants of great environmental concern because of their toxic, mutagenic and carcinogenic properties. Microbial degradation of PAHs is thought to be the major process involved in effective bioremediation of contaminated soils. But PAH removal during bioremediation is often incomplete and residual concentrations after bioremediation are often too high to satisfy the standards for clean soil. These high residual PAH concentrations are usually caused by the limited bioavailability of PAHs which is related to the solubility in water. The CD molecules are ideal hosts for the solubilization of low-polarity guest compounds like PAHs. It has been known that with the increased concentration of HP-β-CD or Me-β-CD, the solubility of the PAHs increased. The solubility of phenanthrene and pyrene was increased by 40–50 times in 5% (w/v) HP-β-CD as well as that of naphthalene, acenaphthalene, anthracene or fluoranthene was enhanced 4–13 times [93]. The enhanced solubility can lead to the enhanced bioavailability and the corresponding faster degradation rates.

The bioremediation of soils contaminated by another pollutant of PCB is also limited by its low bioavailability. RM-β-CD were found to be a potential enhancing agent for bioavailability of PCB in the aerobic treatment process. When the soils, contaminated by about 890–8,500 mg/kg of PCBs, had been treated by 0.5%–1.0% of RM-β-CD for 6 h, a significant depletion of the initial soil ecotoxicity was detected [94]. The RM-β-CD effects generally increased proportionally with its concentration. RM-β-CD was slowly metabolized by the aerobic microorganisms in soil and was found to be favorable in enhancing the occurrence of the indigenous aerobic or the aerobic bioremediation by increasing the bioavailability of PCBs.

The removal of 2,4,6-trinitrotoluene (TNT) from contaminated soil by CDs had been reported, in which Me-β-CDs were used to flush the soil and the soil extract solutions with high TNT loads were further treated by Photo-Fenton [95]. The results showed that the Me-β-CD solution increased the aqueous concentration of TNT in soil extract effluents 2.1 times as much as the concentrations obtained during the water flush of the soil. Photo-Fenton reaction is an efficient process for final disposal of TNT contaminated soil extract solutions. Me-β-CD has a beneficial effect on TNT degradation rates in complex solutions containing high amounts of hydroxyl radical scavengers. Although at last soil extract solution mineralization was not completed, no potential toxic aromatic intermediates were left in the treated solution. These results indicate that the CD-enhanced solubilization of TNT with Photo–Fenton treatment may be a promising approach for TNT contaminated soil remediation.

Recently an *in situ* CD-enhanced combined technology has been introduced for remediation of a site polluted with less biodegradable mixture of aged diesel and engine oils, in which RM-β-CD was used to enhance the biodegradation and solubilization of contaminants, and the importance of the technology monitoring was emphasized [96]. The combined technology consists of the following steps: First, the actual site is demonstrated, then the laboratory experiments and analysis with the contaminated soils originating from the site is carried out. Next, a technology is designed based on the detailed assessment of the contaminated site. During remediation, the technology monitoring has responsibility for characterization of the processes at the whole treated site.

Soil wettability is another important factor besides contamination, which governs water retention and transport processes [97]. Strong interaction of mineral phases with CDs suggests that the properties of the soils may also be seriously affected by CDs, which may have an influence on soil remediation processes. The effect of RM-β-CD on surface and pore properties of common clay minerals such as bentonite, illite and kaolinite had been studied and it was found that the wettability of the soil decreased at low CD concentration and increased again at the highest load. The increased concentration of CDs reduced the effective radius of the soil bed which improved the average force of interaction among soil particles via water layer, despite the simultaneous decrease of adhesion forces of water to the soil.

8.2.4. *Environmental bioassays*

With the development of application technologies of CDs, some newly CD-aided tools have been deployed in sampling, measuring the concentration or testing the effect of contaminants in water or soil. The innovative methods such as the bacterial bioassays with CD-increased sensitivity or the CD-filled absorptive samplers for air and water sampling are utilized in environmental hazard and risk assessment. These new technologies are used for the reduction of the risk of chemical compounds in waters and soils. CD-aided environmental remediation has been introduced for the elimination of organic contaminants from water or the enhancement of the mobility and availability of soil contaminants, through increasing the efficiency of soil remediation by water extraction, chemical oxidation, biodegradation with CDs and their derivatives.

Bioassays are able to simulate and measure actual effects of the contaminants in a certain environment, but the application of bioassays for environmental samples especially for soil is a difficult issue which may result in low statistics and an under estimation of the risk of the environmental sample because of the low reliability and sensitivity caused by strong interactions, matrix effects and non-equilibrium state. Direct contact of the test organism with soil improves this situation, but the restricted bioavailability may cause under estimated adverse effects and risk. Adding CDs could increase bioavailability, bioaccessibility and partition between soil phases and consequently improve the sensitivity of the test by incorporating organic pollutants, which make the possibility of the application of CDs in bioassays. In the evaluation of pentachlorophenol (PCP) in soil, RM-β-CD was used in three bioassay systems of bacterial luminescence-inhibition test with *Vibrio fischeri*, protozoon growth inhibition test with *Tetrahymena pyriformis* and Ames mutagenicity test [98]. The results indicated that RM-β-CD surely has an influence on the fate of the contaminant in soil, and the change tendency induced by RM-β-CD depends on its effective concentration, contact time, the type of test organism and even the complex characteristics. In those cases where the pollution effects were enhanced by RM-β-CD, RM-β-CD is very useful in risk assessment, resulting in a moderate overestimate in the value of environmental risk.

The effects of aging on the availability of pyrene, earthworm (*Eisenia fetida*) accumulation and chemical extraction by exhaustive and nonexhaustive techniques in soil spiked with a range of pyrene levels were measured using both unaged and aged soil samples [98]. The results showed that the amount of pyrene accumulated by earthworms did not remarkably change with aging time at high-dose levels of contaminant while it changed significantly at lower concentrations. It was found that 222 days of aging greatly reduced biological and chemical availability of pyrene. Furthermore, the relationship among earthworm bioaccumulation, HP-β-CD, and organic solvent extraction was investigated in order to find a suitable and rapid approach to predict pyrene bioavailability. Results showed that the mild HP-β-CD extraction might be a better method to predict bioavailability of pyrene in soil.

8.3. Application of CDs in Agriculture

In agriculture, CDs can be applied to delay germination of seeds. Some amylases are inhibited in grain treated with β-CD. β-CD treatment resulted in initial slow growth of the plant but later an improved plant growth was observed yielding a 20%–45% larger harvest than that of the control. Recent developments involve the expression of CGTases in plants. Another application field of CDs is to improve the physico–chemical characteristics of pesticides. CDs can form inclusion complexes with a variety of pesticides which results in an improvement in physical, chemical and biological properties of pesticides, such as the increase in stability of degradable pesticides, the enhancement of bioactivity and the acceleration in degradation of pesticides [99]. The advantages of the use of CD complexes in the pesticide industry have two additional features compared with those in food or pharmaceutical fields. First, the incorporated pesticides have hydrophilic covers, which mean that absorption into the hydrophilic environment in the intestine of a pest is increased, whereas contact absorption through the hydrophobic surface of the insect exoskeleton is decreased. This suggests that these complexes can be used to selectively attack herbivorous pests, while other insects remain unaffected by contact absorption. Second, the incorporated pesticides possess characteristics of timed-release activity.

Butachlor is a chloracetamide herbicide widely used for the control of annual grasses and is highly toxic to many organisms. It has been known that soil adsorption is one of the key factors influencing bioavailability of hydrophobic chemicals. Usually, higher amounts of the herbicide butachlor are necessary to achieve its herbicidal activity due to its low water solubility and high hydrophobicity, hence increasing its environmental risks. The effects of β-CD on solubility and soil adsorption of butachlor had been investigated. Butachlor could form a 1:1 stoichiometric CD complex whose apparent stability constant was 443 L·mol^{-1}, and the corresponding dissolution of butachlor was enhanced significantly. The adsorption of butachlor on soil was reduced with an increase of β-CD concentration because of the formation of the inclusion complex with low adsorption potency. Although the sorption distribution coefficient of complexed butachlor was about 14% of that of the free herbicide, the proportion of the adsorbed amount of complexed butachlor to the total adsorbed amount rose with the increase of β-CD concentration [100]. Thus, the adsorption of inclusion complex cannot be neglected in the presence of high concentration CDs, although its water solubility was much higher than that of the free herbicide. These results indicate that β-CD may be used as a formation additive to improve the solubility of butachlor, reduce its adsorption on soil and increase the availability of butachlor for weeds.

The herbicide norflurazon could be complexed by β-CD with 1:1 stoichiometric ratio, and up to five-fold increase in norflurazon solubility was obtained. Further study developed controlled release and/or protective formulations of norflurazon, which allow a more rational application of norflurazon, diminishing the use of organic solvents and increasing its efficacy [101]. With regard to the pesticide chloramidophos (CP), complexation by β-CD greatly improved its thermal stability and had no adverse effects on its bioefficacy after being evaluated by an *in vitro* acetylcholinesterase (AChE) inhibition assay and an acute aquatic toxicity assay. This also suggested a promising outlook for the complexed CP to be an active ingredient for various formulation products of CP.

RM-β-CD and HP-β-CD also could be used to improve the solubility of 10-undecyn-1-ol. *In vitro* evaluations of the growth inhibition effects of inclusion complex solutions toward *Rosellinia necatrix*, a phytopathogenic fungus were carried out. In comparison with the positive control, the antifungal activity of 10-undecyn-1-ol after adding CD derivatives was obviously improved. RM-β-CD was proven to be more effective compared to HP-β-CD with regard to the reduction of both fungal mycelium-covered area and growth rate constant as well as the bioavailability improvement of 10-undecyn-1-ol [102]. These results suggest potential approaches for production of environmentally friendly 10-undecyn-1-ol kinds of fungicide.

References

1. Szejtli, J (1988). *Cyclodertrin Technology*. Boston: Kluwer Academic Publishers.
2. Szejtli, J (1998). Introduction and general overview of cyclodextrin chemistry. *Chemical Reviews*, 98, 1743.
3. Liu, L, KS Song, XS Li, and QX Guo (2001). Charge-transfer interaction: A driving force for cyclodextrin. Inclusion complexation. *Journal of Inclusion Phenomena and Macrocyclic Chemistry*, 40, 35–39.
4. Szejtli, J (2004). Past, present, and future of cyclodextrin research. *Pure Applied Chemistry*, 76(10), 1825–1845.
5. Li, S, WC Purdy (1992). Cyclodextrins and their applications in analytical chemistry. *Chemical Reviews*, 92, 1457–1470.
6. Mosinger, J, T Viktorie, N Irena and Z Jaroslav (2001). Cyclodextrins in analytical chemistry. *Analytical Letters*, 34(12), 1979–2004.
7. Shpigun, OA, IA Ananieva, NB Yu and EN Shapovalova (2003). Use of cyclodextrins for separation of enantiomers. *Russian Chemical Reviews*, 72, 121035–121054.
8. Dodziuk, H (2006). *Cyclodextrins and Their Complexes: Chemistry, Analytical Methods, Applications*, Germany, p. 507: Wiley VCH. .
9. Juvancz, Z and J Szejtli (2002). The role of cyclodextrins in chiral selective chromatography. *TrAC Trends in Analytical Chemistry*, 21(5), 379–388.
10. Günther, K, P Richter and K Möller (2003). Separation of enantiomers by thin-layer chromatography: An overview. In *Chiral Separations Methods and Protocols*, Methods in Molecular Biology, Vol. 243, pp. 29–59.
11. Hinze, WL, DY Pharr, ZS Fu and WG Burkert (1989). Thin-layer chromatography with urea solubilized beta.-cyclodextrin mobile phases. *Analytical Chemistry*, 61(5), 422–428.
12. Lepri, L, L Boddi, M Del ubba and A Cincinelli (2001). Reversed-phase planar chromatography of some enantiomeric amino acids and oxazolidinones. *Biomedical Chromatography, Special Issue: Chiral Resolution*, 15(3), 196–201.
13. Touchstone, JC (1992). *Practice of Thin Layer Chromatography*. Wiley-Interscience publication.
14. Duncan, JD and DW Armstrong (1990). Normal phase TLC separation of enantiomers using chiral ion interaction agents. *Journal of Liquid Chromatography*, 13(6).

15. Huang, MB, HK Li, GL Li, CT Yan and LP Wang (1996). Planar chromatographic direct separation of some aromatic amino acids and aromatic amino alcohols into enantiomers using cyclodextrin mobile phase additives. *Journal of Chromatography A*, 742, 289–294.
16. Politzer, LR, KT Crago, T Hollin and M Young (1995). TLC of p-Nitroanilines and Their Analogues with Cyclodextrins in the Mobile Phase. *Journal of Chromatographic Science*, 33(6), 316–320.
17. Sasaki, T, M Wakabayashi, T Yamaguchi, Y Kasuga, M Nagatsuma, T Iida and T Nambara (1999). Separation of double conjugates of bile acids by two-dimensional high-performance thin-layer chromatography with tetra-n-butylammonium phosphate and methyl β-cyclodextrin. *Chromatographia*, 49(11–12), 681–685.
18. Momose, T, M Mure, T Iida, J Goto and T Nambara (1998). Method for the separation of the unconjugates and conjugates of chenodeoxycholic acid and deoxycholic acid by two-dimensional reversed-phase thin-layer chromatography with methyl beta-cyclodextrin. *Journal of Chromatography A*, 811(1–2), 171–180.
19. Grinberg, N, G Bicker, P Tway and JA Baiano (1988). Recognition of artifacts occurring in TLC. *Journal of Liquid Chromatography*, 11(15).
20. Berthod, A, W Li and DW Armstrong (1992). Multiple enantioselective retention mechanisms on derivatized cyclodextrin gas chromatographic chiral stationary phases. *Analytical Chemistry*, 64(8), 873–879.
21. Smolková-Keulemansová, E (1982). Cyclodextrins as stationary phases in chromatography. *Journal of Chromatography A*, 251(1), 17–34.
22. Konig, WA, SLM Hagen and R Krebber (1989). Cyclodextrins as chiral stationary phases in capillary gas chromatography, Part IV: Heptakis(2,3,6-tri-O-pentyl)-β-cyclodextrin. *Journal of High Resolution Chromatography*, 12, 35–39.
23. Wenz, G, P Mischnick, R Krebber, M Richters and AK Wilfried (1990). Preparation and characterization of per-O-pentylated cyclodextrins. *Journal of High Resolution Chromatography*, 13(10), 724–728.
24. Ochocka, RJ, D Sybilska, M Asztemborska, J Kowalczyk and J Goronowicz (1999). Approach to direct chiral recognition of some terpenic hydrocarbon constituents of essential oils by gas chromatography systems via α-cyclodextrin complexation. *Journal of Chromatography A*, 543, 171–177.
25. Armstrong, DW, W Li, AM Stalcup, HV Secor, RR Izac and JI Seeman. Capillary gas-chromatographic separation of enantiomers with stable dipentyl-α-, β- and γ-cyclodextrin-derivatized stationary phases.
26. Armstrong, DW, W Li, AM Stalcup, HV Secor, RR Izac and JI Seeman (1990). Capillary gas chromatographic separation of enantiomers with stable dipentyl-α, β- and γ-cyclodextrin-derivatized stationary phases. *Analytica Chimica Acta*, 234, 365–380.
27. A Selection Guide to DEXTM Columns : Stable derivatized cyclodextrin stationary phases for high resolution analyses of optical and positional isomers. In *Chiral Cyclodextrin Capillary GC Columns*. Sigma-Aldrich Co. Available at http://www.sigmaaldrich.com/etc/medialib/docs/Supelco/Bulletin/6504.Par.0001.File.tmp/6504.pdf.
28. Solms, RH (1965). Egli. Harze mit Einschlusshohr äumen von Cyclodextrin-Struktur. *Helvetica Chimica Acta*, 48(6), 1225–1228.

29. Stalcup, AM and KH Gahm (1996). A sulfated cyclodextrin chiral stationary phase for high-performance liquid chromatography. *Analytiacl Chemistry*, 68(8), 1369–1374.

30. Bressolle, F, M Audran, TN Pham and JJ Vallon (1996). Cyclodextrins and enantiomeric separations of drugs by liquid chromatography and capillary electrophoresis: Basic principles and new developments. *Journal of Chromatogrphy B: Biomedical Applications*, 687(2), 303–336.

31. Anan'eva, IA, EN Shapovalova, SA Lopatin, OA Shpigun, VP Varlamov, VA Davankov and DW Armstrong (2002). Use of unmodified and aminated β-cyclodextrins for the separation of enantiomers of amino acid derivatives by high-performance liquid chromatography. *Journal of Analytical Chemistry*, 57(4), 331–337.

32. Chemical identity of the enantioselective stationary phases. In Stationary Phases. Available at http://chiral-separations.com/Stationary phases.

33. Araki, T, Y Kashiwamoto, S Tsunoi, M Tanaka (1999). Preparation and enantiomer separation behavior of selectively methylated γ-cyclodextrin-bonded stationary phases for high-performance liquid chromatography. *Journal of Chromatography A*, 845(1–2), 455–462.

34. Ryu, JW, HS Chang, YK Ko, JC Woo, DW Koo and DW Kim (1999). Direct chiral separation of tryptophan analogues using heptakis(3-O-Methyl)-beta-cyclodextrin-bonded stationary phase in reversed-phase liquid chromatography. *Microchemical Journal*, 63(1).

35. Ciucanu, I and WA König (1994). Immobilization of peralkylated β-cyclodextrin on silica gel for high-performance liquid chromatography. *Journal of Chromatography A*, 685(1), 166–171.

36. Farkas, G, LH Irgens, G Quintero, MD Beeson, A Al-Saeed and G Vigh (1993). Displacement chromatograpy on cyclodextrin silicas, IV. Separation of the Enantiomers of Ibuprofen. *Journal of Chromatography*, 645, 67–74.

37. Li, S and WC Purdy (1991). Liquid chromatographic separation of the enantiomers of dinitrophenyl amino acids using a ß-cyclodextrin-bonded stationary phase. *Journal of Chromatography*, 543, 105–112.

38. Bhushan, R and D Gupta (2006). HPLC resolution of thioridazine enantiomers from pharmaceutical dosage form using cyclodextrin-based chiral stationary phase. *Journal of Chromatography B*, 837, 133–137.

39. Nowakowski, R, PJP Cardot, AW Coleman, E Villard and G Guiochon (1995). Elution mechanisms of cyclodextrins in reversed phase chromatography. *Analytical Chemistry*, 67(2), 259–266.

40. Chen, D, S Jiang, Y Chen and Y Hu (2004). HPLC determination of sertraline in bulk drug, tablets and capsules using hydroxypropyl-β-cyclodextrin as mobile phase additive. *Journal of Pharmaceutical and Biomedical Analysis*, 34, 239–245.

41. Chen, CY, CH Lin and JH Yang (2005). Use of chemically bonded β-cyclodextrin silica stationary phase for liquid chromato graphic separation of structural isomers. *Journal of the Chinese Chemical Society*, 52(4), 753–758.

42. Shimada, K, T Oe and M Suzuki (1991). Effect of derivatization of steroids on their retention behaviour in inclusion chromatography using cyclodextrin as a mobile phase additive. *Journal of Chromatography A*, 558(1), 306–331.

43. Dall'asta, C, G Ingletto, R Corradini, G Galaverna and R Marchelli (2003). Fluorescence enhancement of aflatoxins using native and substituted cyclodextrins. *Journal of Inclusion Phenomena and Macrocyclic Chemistry*, 45(3–4), 257–263.

44. Hamilton, LM, CT Kelly and WM Fogarty (2000). Review: Cyclodextrins and their interaction with amylolytic enzymes. *Enzyme and Microbial Technology*, 26(8), 561–567.

45. Kriegshäuser, G and W Liebl (2000). Pullulanase from the hyperthermophilic bacterium Thermotoga maritima: Purification by beta-cyclodextrin affinity chromatography. *Journal of Chromatography B: Biomedical Science and Applications*, 737(1–2), 245–251.

46. Adachi, M, B Mikami, T Katsube and S Utsumi (1988). Crystal structure of recombinant soybean β-amylase complexed with β-cyclodextrin. *The Journal of Biological Chemistry*, 273, 19859–19865.

47. Yoshigi, N, H Sahara and S Koshino (1994). Role of the C-terminal region of β-amylase from barley. *The Journal of Biochemistry*, 117(1), 63–67.

48. Li, B, JC Servaites and DR Geiger (1992). Characterization and subcellular localization of debranching enzyme and endoamylase from leaves of sugar beet. *Plant Physiology*, 98(4), 1277–1284.

49. Wisessing, A, A Engkagul, A Wongpiyasatid and K Chuwongkomon (2008). Purification and Characterization of *C. maculatus* α-amylase. *Kasetsart Journal: Natural Science*, 42, 240–244.

50. Joao Luiz Pinheiro Bastos, José Tarquinio prisco and Enéas. Gomes Filho. (1994). Purification and characterization of a cotyledonary α-amylase from cowpea seedlings. *Revista Brasileira de Fisiologia Vegetal*, 6(1), 33–39.

51. Fukuda, K, Y Teramoto, M Goto, J Sakamoto, S Mitsuiki and S Hayashida (1992). Specific inhibition by cyclodextrins of raw starch digestion by fungal glucoamylase. *Bioscience, Biotechnology and Biochemistry*, 56(4), 556–559.

52. Weisz, PB and PA Yardley. Cyclodextrin polymers and cyclodextrins immobilized on a solid surface. US Patent Publication, 5262404.

53. Witsuba, D and V Volker schurig (2009). The separation of enantiomers on modified cyclodextrins by capillary electrochromatography (CEC), LCGC Europe, 22(2).

54. Kodama, S, A Yamamoto and A Matsunaga (1998). Direct chiral resolution of pantothenic acid using 2-hydroxypropyl-β-cyclodextrin in capillary electrophoresis. *Journal of Chromatography A*, 811(1–2), 269–273.

55. Maruszak, W, MG Schmid, G Gübitz and E Ekiert (2004). Marek trojanowicz separation of enantiomers by capillary electrophoresis using cyclodextrins. *Methods in Molecular Biology*, 243, 275–289.

56. de Boer, T, RA de Zeeuw, GJ de Jong and K Ensing (2000). Recent innovations in the use of charged cyclodextrins in capillary electrophoresis for chiral separations in pharmaceutical analysis. *Electrophoresis*, 21(15), 3220–3239.

57. Pumeral, M, R Matalovál, I Jelínek, J Jindřich and J Jůza (2001). Comparison of association constants of cyclodextrins and their tert-butyl derivatives with halogenbenzoic acids and acridine derivatives. *Molecules*, 6, 221–229.

58. Juvancz, Z, I Urmös and I Klebovich (1996). Capillary electrophoretic separation of clomiphene isomers using various cyclodextrins as additives. *Capillary Electrophoresis*, 3(4), 181–189.

59. Ong, CP, CL Ng, HK Lee and SFY Li (1991). Separation of water- and fat-soluble vitamins by micellar electrokinetic chromatography. *Journal of Chromatography A*, 547, 419–428.

60. Razak, JL, HJ Doyen and CE Lunte (2003). Cyclodextrin-modified micellar electrokinetic chromatography for the analysis of Esterom, a topical product consisting of hydrolyzed benzoylecgonine in propylene glycol. *Electrophoresis*, 24(11), 1764–1769.

61. Yu, L. (2004). Spectroscopic studies on molecular recognition of modified cyclodextrins. *Current Organic Chemistry*, 8, 3546.

62. Wehry, EL. Molecular fluorescence and phosphorescence spectrometry. In *Handbook of Instrumental Techniques for Analytical Chemistry*, Chapter 25, pp. 507–539.

63. Szente, L and J Szejtli (1998). Non-chromatographic analytical uses of cyclodextrins. *Analyst*, 123, 735–741.

64. León-Rodríguez, LMD and DA Basuil-Tobias (2005). Testing the possibility of using UV–vis spectrophotometric techniques to determine non-absorbing analytes by inclusion complex competition in cyclodextrins. *Analytica Chimica Acta*, 543(1–2), 282–290.

65. Tang, B, L Ma, HY Wang and GY Zhang (2002). Study on the supramolecular interaction of curcumin and beta-cyclodextrin by spectrophotometry and its analytical application. *Journal of Agriculture Food Chemistry*, 50(6), 1355–1361.

66. Eastwood, D (1985). Advances in luminescence spectroscopy: A symposium. ASTM Committee E-13 on Molecular Spectroscopy.

67. Wehry, EL. Molecular fluorescence and phosphorescence spectrometry. In *Handbook of Instrumental Techniques for Analytical Chemistry*, Chapter 25, pp. 507–539.

68. Aicart, E and E Junquera (2003). Complex formation between purine derivatives and cyclodextrins: A fluorescence spectroscopy study. *Journal of Inclusion Phenomena and Macrocyclic Chemistry*, 47(3–4), 161–165.

69. Wagner, BD (1998). The fluorescence enhancement of 1-anilinonaphthalene-8-sulfonate (ANS) by modified β-cyclodextrins. *Journal of Photochemistry and Photobiology A Chemistry*, 114(2), 151–157.

70. Kano, K, S Hashimoto, A Imai and T Ogawa (1984). Three-component complexes of cyclodextrins. Exciplex formation in cyclodextrin cavity. *Journal of Inclusion Phenomena and Macrocyclic Chemistry*, 2(3–4), 737–746.

71. Maragos, CM, M Appell, V Lippolis and AL Visconti (2008). Use of cyclodextrins as modifiers of fluorescence in the detection of mycotoxins. *Food Additives & Contaminants: Part A*, 25(2), 164–171.

72. Richmond, MD and RJ Hurtubise (1989). Analytical characteristics of beta.-cyclodextrin/salt mixtures in room-temperature solid-surface luminescence analysis. *Analytical Chemistry*, 61(23), 2643–2647.

73. Kathuria, A, A Gupta, N Priya, P Singh, HG Raj, AK Prasad, VS Parmar and SK Sharma (2009). Specificities of calreticulin transacetylase to acetoxy derivatives of 3-alkyl-4-methylcoumarins: Effect on the activation of nitric oxide synthase. *Bioorganic & Medicinal Chemistry*, 17(4), 1550–1556.

74. Deumié, JM, KLP Kubát and DM Wagnerová (2000). Supramolecular sensitizer: Complexation of meso-tetrakis (4-sulfonatophenyl) porphyrin with 2-hydroxypropyl-cyclodextrins. *Journal of Photochemistry and Photobiology A: Chemistry*, 130(1), 13–20.

75. Lee, S, D Yi and S Jung (2004). NMR spectroscopic analysis on the chiral recognition of noradrenaline by β-cyclodextrin (β-CD) and Carboxymethyl-β-cyclodextrin (CM-β-CD). *Bulletin of the Korean Chemical Society*, 25(2), 216–220.
76. Thomas, JW and DW James (2003). Chiral reagents for the determination of enantiomeric excess and absolute configuration using NMR spectroscopy. *Chirality*, 15(3), 256–270.
77. Susan, EB, HC John, FL Stephen and R Daniel (1991). Coghlan and Christopher J. Easton. Chiral molecular recognition: A 19F nuclear magnetic resonance study of the diastereoisomer inclusion complexes formed between fluorinated amino acid derivatives and α-cyclodextrin in aqueous solution. *Journal of the Chemical Society, Faraday Transactions*, 87, 2699–2703.
78. Radi, AE and S Eissa (2010). Electrochemistry of cyclodextrin inclusion complexes of pharmaceutical compounds. *The Open Chemical and Biomedical Methods Journal*, 3, 74–85.
79. Pospíšil, L, RMPC Trsková and R Fuoco (1998). Inclusion complexes of atrazine with α-, β- and γ-cyclodextrins. Evidence by polarographic kinetic currents. *Journal of Inclusion Phenomena and Macrocyclic Chemistry*, 31(1), 57–70.
80. Cardona, CM, TD McCarley and AE Kaifer (2000). Synthesis, electrochemistry, and interactions with β-cyclodextrin of dendrimers containing a single ferrocene subunit located "off-center". *The Journal of Organic Chemistry*, 65(6), 1857–1864.
81. Fakayode, SO, M Lowry, A Kristin, KA Fletcher, X Huang, AM Powe and IM Warner (2007). Cyclodextrins host- guest chemistry in analytical and environmental chemistry, *Current Analytical Chemistry*, 3(3), 171–181(11).
82. Fenyvesi, E, K Gruiz, S Verstichel, BD Wilde, L Leitgib K Csabai and N Szaniszlo (2005). Biodegradation of cyclodextrins in soil. *Chemosphere*, 60, 1001–1008.
83. Blach, P, S Fourmentin, D Landy and F Cazier (2008). Gheorghe surpateanu a cyclodextrins: A new efficient absorbent to treat waste gas streams. *Chemosphere*, 70, 374–380.
84. Fourmentin, S, D Landy, P Blach, E Piat, G Surpateanu and G Surpateanu (2006). Cyclodextrins: A potential absorbent for VOC abatement. *Global NEST Journal*, 8(3), 324–329.
85. Favier, IM, D Baudelet and S Fourmentin (2011). VOC trapping by new crosslinked cyclodextrin polymers. *Journal of Inclusion Phenomena and Macrocyclic Chemistry*, 69, 433–437.
86. Szente, L, E Fenyvesi and J Sefszejtli (1999). Entrapment of iodine with cyclodextrins: Potential application of cyclodextrins in nuclear waste management. *Environmental Science and Technology*, 33, 4495–4498.
87. Zhang, A and W Liu (2008). Inclusion effect of alpha-cyclodextrin on chemical degradation of malathionin water. *Archives of Environmental Contamination and Toxicology*, 54, 355–362.
88. Crini, G (2005). Recent developments in polysaccharide-based materials used as adsorbents in wastewater treatment. *Progress in Polymer Science*, 30, 38–70.
89. Crini, G (2008). Kinetic and equilibrium studies on the removal of cationic dyes from aqueous solution by adsorption onto a cyclodextrin polymer. *Dyes and Pigments*, 77, 415–426.

90. Eti, BT, Y Mastai and K Landfester (2010). Miniemulsion polymerization of cyclodextrin nanospheres for water purification from organic pollutants. *European Polymer Journal*, 46, 1671–1678.

91. Zhang, X, F Wu and NH Deng (2011). Efficient photodegradation of dyes using light-induced self assembly TiO_2/β-cyclodextrin hybrid nanoparticles under visible light irradiation. *Journal of Hazardous Materials*, 185, 117–123.

92. Bardi1, L, A Mattei, S Steffan and M Marzona (2000). Hydrocarbon degradation by a soil microbial population with b-cyclodextrin as surfactant to enhance bioavailability. *Enzyme and Microbial Technology*, 27, 709–713.

93. Ramsay, JA, K Robertson, G Loon, N Acay and BA Ramsay (2005). Enhancement of PAH biomineralization rates by cyclodextrins under Fe(III)-reducing conditions. *Chemosphere*, 61, 733–740.

94. Chen, Y, X Tang, SA Cheema, W Liu and C Shen (2010). Beta-cyclodextrin enhanced phytoremediation of aged PCBs-contaminated soil from e-waste recycling area. *Journal of Environmental Monitoring*, 12(7), 1482–1489.

95. Sheremata, TW and J Hawari (2000). Cyclodextrins for desorption and solubilization of 2,4,6-trinitrotoluene and Its metabolites from soil. *Environmental Science and Technology*, 34(16), 3462–3468.

96. Leitgiba, L, K Gruiza, É Fenyvesib, G Balogha and A Murányic (2008). Development of an innovative soil remediation: "Cyclodextrin-enhanced combined technology, *Science of the Total Environment*, 392, 12–21.

97. Józefaciuk, G, M Hajnos and A Muranyi (2002). Influence of a cyclodextrin on soil wettability. *International Agrophysics*, 16, 111–118.

98. Hajdu, C, K Gruiz, É Fenyvesi and Z Nagy (2011). Application of cyclodextrins in environmental bioassays for soil. *Journal of Inclusion Phenomena and Macrocyclic Chemistry*, 70(3–4), 307–313.

99. Khan, MI, SA Cheema, C Shen, C Zhang, X Tang, Z Malik, X Chen and Y Chen (2011). Assessment of pyrene bioavailability in soil by mild hydroxypropyl-β-cyclodextrin extraction. *Archives of Environmental Contamination and Toxicology*, 60(1), 107–115.

100. Martin Del Valle, EM (2003). Cyclodextrins and their uses: A review. *Process Biochemistry*.

101. Bian, H, J Chen, X Cai, P Liu, H Liu, X Qiao and L Hua (2009). Inclusion complex of butachlor with β-cyclodextrin: Characterization, solubility, and speciation-dependent adsorption. *Journal of Agricultural and Food Chemistry*, 57, 7453–7458.

102. Gruiz, K, M Molnár, E Fenyvesi, C Cs. Hajdu, Á Atkári and K Barkács (2011). Cyclodextrins in innovative engineering tools for risk-based environmental management. *Journal of Inclusion Phenomena and Macrocyclic Chemistry*, 70(3–4), 299–306.

103. Neoh, TL, T Tanimoto, S Ikefuji, H Yoshii and T Furuta (2008). Improvement of antifungal activity of 10-undecyn-1-ol by inclusion complexation with cyclodextrin derivative. *Journal of Agriculture Food Chemistry*, 56(10), 3699–3705.

INDEX

.

www.ingramcontent.com/pod-product-compliance
Lightning Source LLC
Chambersburg PA
CBHW050547190326
41458CB00007B/1945